Stories from the Wreckage

STORIES
from the
WRECKAGE

A GREAT LAKES MARITIME HISTORY
INSPIRED BY SHIPWRECKS

JOHN ODIN JENSEN

Featuring the Photography and Fieldwork of the
Wisconsin Historical Society Maritime Preservation
and Archaeology Program

WISCONSIN HISTORICAL SOCIETY PRESS

Published by the Wisconsin Historical Society Press

Publishers since 1855

The Wisconsin Historical Society helps people connect to the past by collecting, preserving, and sharing stories. Founded in 1846, the Society is one of the nation's finest historical institutions.

Join the Wisconsin Historical Society:
wisconsinhistory.org/membership

Publication of this book was made possible in part by funding from the Wisconsin Department of Transportation.

Photographs identified with WHI or WHS are from the Society's collections; address requests to reproduce these photos to the Visual Materials Archivist at the Wisconsin Historical Society, 816 State Street, Madison, WI 53706.

The following cover and front matter photos are by Tamara Thomsen: front cover, *Home* shipwreck, WHI IMAGE ID 119678; page ii, *Australasia* shipwreck, WHI IMAGE ID 141640; page vi, Wisconsin Historical Society diver, WHI IMAGE ID 141641; and back cover, Wisconsin Historical Society diver, WHI IMAGE ID 140132, and historical photo of the *Australasia*, C. Patrick Labadie Collection / Thunder Bay National Marine Sanctuary, Alpena, MI.

Printed in the United States of America

Designed by Percolator

23 22 21 20 19   1 2 3 4 5

Library of Congress Cataloging-in-Publication Data

Names: Jensen, John Odin, author.
Title: Stories from the wreckage : a Great Lakes maritime history inspired
    by shipwrecks / John Odin Jensen.
Description: Madison, WI : Wisconsin Historical Society Press, 2019. |
    Includes bibliographical references and index. |
Identifiers: LCCN 2018049438 (print) | LCCN 2018058801 (ebook) |
    ISBN 9780870209031 (ebook) | ISBN 9780870209024 (paperback)
Subjects: LCSH: Shipwrecks—Great Lakes (North America)—History. |
    Navigation—Great Lakes (North America)—History. | Great Lakes
    Region (North America)—History. | BISAC: TRANSPORTATION /
    Ships & Shipbuilding / History. | HISTORY / United States / State &
    Local / Midwest (IA, IL, IN, KS, MI, MN, MO, ND, NE, OH, SD, WI). |
    TECHNOLOGY & ENGINEERING / Marine & Naval.
Classification: LCC G525 (ebook) | LCC G525 .J47 2019 (print) |
    DDC 917.704—dc23

♾ The paper used in this publication meets the minimum requirements of the American National Standard for Information Sciences—Permanence of Paper for Printed Library Materials, ANSI Z39.48-1992.

To my father, Captain James Nordland Jensen,
and all my other shipmates from Alaska and the
Great Lakes who have sailed on the eternal voyage
but taught me so much before they departed.

CONTENTS

The famous schooner *Rouse Simmons* foundered on November 23, 1912, in Lake Michigan near Two Rivers, Wisconsin, while bound for Chicago with a cargo of Christmas trees. An estimated sixteen crew members and passengers lost their lives. Local divers place a Christmas tree on the bow once a year. TAMARA THOMSEN/ WHI IMAGE ID 120451

Introduction

SHIPWRECKS AND GREAT LAKES HISTORY

Heartbreaking and often tragic events, shipwrecks have occurred since people built the first rafts from reeds, sticks, or animal skins. Although common in history, shipwrecks affect us in deep psychological and even mystical ways. We do not build museums devoted to automobile accidents or airplane crashes. Yet maritime museums feature shipwreck stories wherever mariners in large numbers faced angry seas and deadly coasts. As narrative events, shipwrecks feature high drama with epic struggles against elemental forces of nature, tales of individual heroism or cowardice, disaster or deliverance, life or death. This book includes several such stories, some of them highly detailed.

However, rather than a history of shipwrecks, this book is a maritime history inspired by shipwrecks. Beyond the final disaster or abandonment, each shipwreck is also the consequence and the convergence point of larger patterns of historical events, factors, processes, and social networks. In some respects, shipwrecks are like the individual blocks that make up the complex patchwork quilt of Great Lakes history. While many blocks look similar and others unique, the completed quilt projects a larger design illustrated by the relationships of its many parts. In this book, I have approached Wisconsin shipwrecks as component blocks in a larger unfinished historical patchwork, one that illustrates the contributions of the Great Lakes to the American maritime experience, the influences of Atlantic maritime culture on midwestern history, and the economic and technological changes of industrialization that slowly separated Great Lakes maritime technology and culture from its oceanic origins. This is a shipwrecks book, but one that also explores the personal ambitions, cultural influences, and innovations of a collection of shipbuilders, ship owners, and entrepreneurial captains representative of the type who helped to transform the Great Lakes from an isolated western maritime frontier to an industrial maritime frontier. Among their actions were the building and the wrecking of thousands of ships.

The foundations of this book are the more than six hundred wooden ships lost or abandoned in the Great Lakes waters of Wisconsin and the three decades of marine archaeology work carried out by the Wisconsin Historical Society, a state agency headquartered in Madison, Wisconsin. While important underwater archaeological studies have taken place throughout the Great Lakes region, Wisconsin is notable for the consistency of its pubic efforts to study and preserve its historic shipwrecks. For three decades, the Wisconsin Historical Society has spearheaded a systematic and unbroken program of maritime and underwater historic preservation protection and public outreach. I began my career as a historian and underwater archaeologist during the formative years of this program and, while I have since moved on, I have retained a strong professional connection and personal commitment to its public mission. This book is an effort to share what I have learned from the study of Wisconsin shipwrecks about the Great Lakes region and the wider maritime world as it developed between 1820 and 1920.

The stories that unfold in the following chapters read principally as traditional historical narratives; however, they emerged through research that combined historical sources with questions, insights, and data inspired and provided by the archaeological record of shipwrecks. When the Wisconsin Historical Society began the active protection and study of historic shipwrecks in the late 1980s, the body of scholarship work on Great Lakes ship design and shipbuilding methods was thin. For most of the nineteenth century, wooden shipbuilders did not work from formal ship plans. Their work was based on collective knowledge, skills, and practices that had evolved in Atlantic maritime communities over several centuries. In spite of this lack of formal design work, shipbuilding innovations, small and large, were the norm on the Great Lakes during this period. They were tested, refined, accepted, or abandoned through the observed performance of the specific ships and in response to rapidly changing conditions of a westward moving and industrializing North America. The proof was in the pudding and not the recipe—and while the recipes might never have been recorded, the end results can still be found underwater.

On its own, this archaeological record cannot explain the human thoughts and motivations responsible for building, operating, and wrecking a ship. But analyzing the physical remains of a wreck in combination with printed historical and visual sources can result in more comprehensive, or at least more broadly connected, interpretations of the past.

The shipwrecks included in this book encompass a relatively brief but crucial period in American and Great Lakes regional history. The oldest of the principal shipwrecks described archaeologically, the magnificent sidewheel steamer *Niagara*, was launched in 1846 and wrecked ten years later. The most recent of the shipwrecks (although not the youngest ship), the wooden bulk carrier *Frank O'Connor*, was built in 1892 and burned off Cana Island in Door County in 1919. These seventy-three years cover part or all of the working lives of nearly every one of Wisconsin's historic Great Lakes shipwrecks. Of the approximately thirty-three ships built before 1846 listed in a recent version of Wisconsin's shipwreck inventory, all but three remained afloat and working during some part of this period.

The oldest ship in the Wisconsin inventory (leaving out the French *Griffon* of 1679) was built in the early 1820s. Between that time and the early 1880s, maritime influences reshaped the cultural and economic landscape of the Great Lakes region to a degree that is difficult to comprehend today. The number of ships transiting Wisconsin's coasts and arriving at its harbors every year grew from a few in number to the many thousands. Small sailing vessels carried fur, fish, and supplies of every kind. Early steamers brought settlers by the tens of thousands. By the 1840s, new schooners of increasing dimensions carried grain grown on frontier farms and lumber from the forests of Michigan, Wisconsin, and Minnesota. By the 1880s, even larger schooners and wooden steam-powered bulk freighters carried iron ore eastward in staggering quantities, returning with coal to heat cities, power factories, and run trains.

By the late nineteenth and early twentieth centuries, heavy industry dominated lake transport in volume and in the size of ships employed. Some older schooners remained in service as did a significant number of older wooden bulk carriers. These vessels remained profitable for owners because of their functional designs and low acquisition costs. Sail and steam lumber carriers remained common until the commercial depletion of the northern Great Lakes forests in the second decade of the twentieth century.

The study of Wisconsin shipwrecks offers a microcosm of a larger regional history and reveals how the personal or cultural relationships between midwestern people and the Great Lakes changed over time. During the early and middle decades of the nineteenth century, the people of Wisconsin and the other Great Lakes states had more intimate and direct connections to the ships and the sea (inland or ocean) than they had at the end of the century. During that period significant portions of the nonfarming population made all or part of their livings working on boats or ships or in shore-based maritime trades such as shipbuilding, blacksmithing, and long-shoring. Other critical areas of the economy, such as fur, agriculture, commercial sales, manufacturing, and lumber, also depended on ships and mariners of one type or another.

This era of regional maritime cultural involvement peaked in the early 1870s, when thousands of commercial watercraft of different types and ages plied the lakes and employed tens of thousands of people. The tonnage of cargo continued to grow well into the twentieth century, but this period marked the climax of the Great Lakes as a maritime cultural system. Beginning in the 1880s, industrialization, urbanization, and technological advances spurred decades of rapid growth in the volume and economic value of Great Lakes shipping. At the same time, however, the maritime dimensions of midwestern culture and the regional identity characteristics of its frontier period began a decline that has continued into the twenty-first century. The historical facts and processes of this transformation are clearly expressed in the archaeological and historical record of Wisconsin's shipwrecks.

Maritime cultural influences in Wisconsin were at first concentrated in Milwaukee and thriving communities along the Lake Michigan shore. Even the rural farmers of the earlier days often had at least one strong maritime memory. European immigrants coming to Wisconsin required a transatlantic voyage and, before the 1850s, a journey across one or more of the Great Lakes. Some of these people, such as the passengers on the ill-fated propeller steamer *Phoenix* that burned near Sheboygan in 1847, braved the Atlantic only to perish just miles from the Wisconsin shore.

The Great Lakes of the nineteenth century were the product of an Atlantic maritime culture and vision. From the earliest French explorers, the most enterprising and visionary of the Great Lakes people sought to forge geographic and economic links with the markets and ports of the Atlantic Ocean to match the long-standing cultural ones. Commercial ships transited between lake and ocean more than a century before the opening of the St. Lawrence Seaway, a fact too often forgotten. By the mid-1850s, the completion of new deep-water canals made it possible for moderate-sized lake vessels to voyage between all of the Great Lakes and out to the ocean.

More important in establishing a vibrant Atlantic maritime culture on the Great Lakes during the 1800s were the movements between ocean and lake of shipbuilders, maritime artisans, and mariners. These people brought Atlantic maritime technologies, skills, legal and business practices, and capital to the frontier and applied them to the design, construction, and operation of ships.

During much of the nineteenth century, the Great Lakes region was part of a larger, expanding western maritime frontier. As other Americans and recent immigrants looked westward toward the frontier for new opportunities, so too did maritime people. Their nautical skills and business acumen proved useful in lands filled with ship timber, good places to build vessels, and expanding markets for the transport of people, foodstuffs, finished goods, and, most significantly, raw materials. Ships held a central place in the economic exchange, and those who reached the middle and upper echelons in the maritime world of the Great Lakes gained social and economic status and, in some cases, significant political power. The most successful were shrewd entrepreneurs who extracted wealth from abundant natural resources and channeled it into railroads, banks, and heavy industry. As histories of Wisconsin shipwrecks reveal, however, many failed.

Characterized by innovation and rapid change during the nineteenth century, the technology of Great Lakes transportation became stable, and in some ways almost static, during the twentieth century. Large steel bulk carriers such as the *Edmund Fitzgerald*, launched in 1958, looked much like those launched five decades earlier and closely resembled the smaller wood bulk carriers constructed in the 1870s and 1880s. Not until the late 1960s, with the launching of the first "thousand footers," did a different class of bulk carrier appear on the lakes. The longevity of some of the early steel freighters is astonishing. The *J. B. Ford*, built in 1903 as the *Edwin F. Holmes*, sailed actively until 1985 and remained in service as a storage vessel until 2006. The ship was finally scrapped in 2015. The *St. Mary's Challenger*, built in 1906 as the *William P. Snyder*, operated as a steam bulk freighter for 107 years before conversion to an articulated barge in 2014. In this configuration, the *St. Mary's Challenger* was still in active service in 2017.

The study of shipwrecks reveals that during the nineteenth century, change and innovation, not technological stability, were the order of the day. The longevity of wooden vessels and of individual ships suggests the extraordinary efficiency and technological prowess achieved by Great Lakes shipbuilders at the close of the era of wooden ships. Out of the chaos of the western and industrial frontiers emerged ships so perfectly suited to the region and its economy as to require little fundamental change.

The archaeology of Wisconsin shipwrecks provides abundant physical evidence of the culture of technological innovation that characterized the leading Great Lakes shipbuilders throughout the nineteenth century. Change in the design of ships and in the methods of their construction came in waves of varying amplitude. While schooners, with nimble and

The *Edmund Fitzgerald*, launched in 1958, retained the basic shape of the Great Lakes bulk steamer pioneered with the *R. J. Hackett* in 1869, but carried more than twenty times the cargo. The *Fitzgerald* foundered with all hands while carrying iron ore in Lake Superior in November 1975. KENNETH THRO COLLECTION, UNIVERSITY OF WISCONSIN-SUPERIOR

The steel *William P. Snyder*, built in 1903 and later renamed the *St. Mary's Challenger*, carried cargo under its own power for more than a century, a longevity unmatched among the world's industrial freighters. CHRIS WINTERS, DISCOVERY WORLD

easy-to-handle fore and aft sailing rigs, dominated the Great Lakes' wind-powered fleets throughout the wooden age, a schooner of the 1820s, such as the 60-ton *Eclipse*, was a very different vessel from the lumber schooner *Lumberman* built in 1862. Beyond their basic sailing rig, the two schooners had little in common with the 700-ton schooner *Lucerne*, built in 1873.

Sailing craft have existed for thousands of years, but practical steam navigation changed the essence of water transport and created the impetus for major changes in ship design. The heavily forested midwestern maritime frontier, with long rivers and huge lakes, provided the perfect environment for steam. By the early 1840s, Great Lakes ship designers such as George Washington Jones and Jacob Banta were launching the largest steamboats afloat anywhere in the world. Great Lakes builders embraced the Ericsson propeller (the forerunner of the modern ship's propeller) as practical technology before builders elsewhere in the United States recognized its merit.[1] Within four years of installing the first Ericsson propeller on the Great Lakes on *Vandalia* in 1841, builders had developed a new and distinctive category of screw-propelled Great Lakes craft. Although George Washington Jones's 1845 *Phoenix* burned near Sheboygan with terrible loss of life in 1847, in general, propellers were at least as safe as and more economical to operate than the older sidewheel steamers.

The region's maritime creativity continued with the development of a fleet of distinctive steam-powered bulk carriers during the late 1860s and early 1870s. These included the *R. J. Hackett*, built in 1869 and wrecked

The sleek grain-carrying schooner *Lucerne*, under full sail when new in 1873. The vessel wrecked with all hands while carrying iron ore in the Apostle Islands area in November 1886. ALPENA COUNTY GEORGE N. FLETCHER PUBLIC LIBRARY

on Whaleback Shoals in Green Bay in 1905, and the *Selah Chamberlain*, built in Cleveland in 1873 and sunk in a collision off of Sheboygan in 1886. A largely forgotten achievement for this period is the early application of steam power to commercial fishing in such vessels as the *T. H. Camp*, built in Lake Ontario in 1876 and lost three decades later in the Apostle Islands. Outside of the Great Lakes, the American fishing industry did not add power to their vessels until the early twentieth century.

During the 1880s and 1890s, Great Lakes builders such as James Davidson of West Bay City, Michigan, pushed the size of wooden cargo ships to their ultimate length, more than three hundred feet, by combining high-quality regional oak and pine with the careful integration of iron and steel strapping. Possessing advanced mechanical systems such as triple-expansion engines and steam-powered steering, yet economical to build, repair, and adapt, the modern-looking large wooden bulk carriers embodied the emerging industrial economy in the Great Lakes region. In the strategic use of the region's dwindling supplies of old-growth timber and in the personal histories of their builders, they offer the final glimpses of the wooden age on the Great Lakes. The Wisconsin wrecks of the Davidson bulk carriers *Australasia*, *City of Glasgow*, *Frank O'Connor*, *Appomattox*, and *Pretoria* provide an appropriate end point in that their methods of construction, financing, and operation mark the limits of individual entrepreneurship in the industrial Great Lakes and the final transformation to a century of corporate fleets and steel ships.

For a historian of the nineteenth century, Great Lakes ships are extraordinarily sensitive indicators of economic, environmental, and technological change. Understanding a shipwreck begins not with its wrecking but in rediscovering the context of its construction. Building a ship is a deliberate act, the collective product of thousands of small and large decisions. By uncovering these decisions, one learns, with stark clarity, how maritime people understood their natural and economic environments and the opportunities they presented. With the original context and purpose of a ship firmly established, one can follow the ship through its career and observe the pressures it operated under and the methods employed and adaptations made to keep it afloat and profitable during periods of nearly constant change. Only then can the vessel's wrecking and the full significance of its archaeological remains be understood.

The constant pace of change in Great Lakes ships and shipping, coupled with the versatility and longevity of some the individual vessels involved, makes the organization of this book challenging. Distinguishing the features

Lake Superior fishermen unload their catches onto the steamer *T. H. Camp*. COURTESY OF THE WHS MARITIME PRESERVATION AND ARCHAEOLOGY PROGRAM

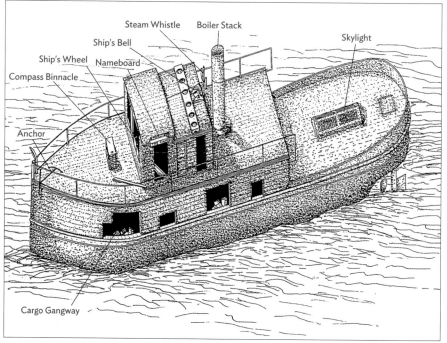

A drawing by artist John Dobbs showing the wreck of the *T. H. Camp* was based on the documentation of a team of avocational underwater archaeologists led by Ken Merryman. JOHN DOBBS AND KEN MERRYMAN

Steam Whistle   Boiler Stack

Ship's Bell   Skylight

Ship's Wheel   Nameboard

Compass Binnacle

Anchor

Cargo Gangway

of the maritime frontier era evident in the archaeological remains of the 1846 steamer *Niagara*, the 1892 wooden bulk steamer *Frank O'Connor*, or the 1862 schooner *Lumberman* from the 1871 *Lucerne,* is relatively easy. Differentiating individual canal schooners such as the *Kate Kelly* that were built from the late 1840s through the early 1870s can be a more difficult task, complicated by the large number built and the survival of so many into the industrial era.

Throughout the mid-1800s on the Great Lakes, the schooner dominated in the number and tonnage of vessels built and in service. A near-perfect workhorse for hauling the low-value bulk cargoes of agricultural products and lumber produced as the frontier developed, schooners also provided the same essential services as during the early decades of industrialization, carrying iron ore and coal. Because these vessels were relatively inexpensive to build, maintain, and operate, and yet were extremely profitable, Great Lakes shipwrights built thousands of them before 1880. Hundreds wrecked in Wisconsin. A bastion of the professional Atlantic sailors and of independent entrepreneur seamen, schooners, and their associated histories illustrate critical dimensions of maritime life during a golden period of commercial sail that ended in economic depression with the Panic of 1873. Their stories also highlight the dark times of competition, declining freight rates, displacement by more efficient industrial ships, and the high rates of shipwreck and death that followed.

This book looks closely at two schooners from this period. The first, aptly named the *Lumberman*, was built in western Michigan in 1862 to carry lumber, and its unusual two-centerboard design illustrates important facets of the Great Lakes forest frontier and the growth of Chicago into the greatest lumber market in the world. The history revealed through the *Lumberman* illustrates intersections of forest and maritime frontier enterprise and innovation that have been largely overlooked by historians. The second schooner, the *Kate Kelly*, is one of hundreds of large schooners designed to transit the Welland Canal between Lakes Erie and Ontario. The ship's general operational history and its design are representative of the important canal schooner and the associated trades it was built to serve. Although representative of its class, the final years of the *Kate Kelly*'s service and its loss with all hands in 1895 contribute also to a larger understanding of the Great Lakes during the wooden age. The career of Hartley Hatch, the ship's final owner and captain, reinforces the fact that Great Lakes ships and mariners were part of an international maritime system that included Canadian waters and extended thousands of miles across the North and

South Atlantic, the Pacific, and even the Indian Ocean. Hatch's story is also a case study of technological innovation, personal independence, and resistance to maritime industrialization at the close of the nineteenth century.

Rather than focusing on a single shipwreck, chapters eight through ten examine the transformation of the Great Lakes from the 1850s through the early decades of the twentieth century through the career, ships, and shipwrecks associated with Captain James Davidson. A cosmopolitan seafarer, innovative shipbuilder, and extraordinarily successful maritime entrepreneur, Davidson's life and times capture decades of rapid changes to the Great Lakes and their consequences. Working the Buffalo waterfront as an orphan boy in the early 1850s, Davidson participated directly in the closing of the early maritime frontier period on the Great Lakes, sailed the Atlantic and Indian Oceans, became a captain and ship owner in the mid-1860s, and beginning in the 1870s became the best-known and, ultimately, the final builder of large wooden ships on the Great Lakes. If success is measured in the acquisition of personal wealth and the minimizing of human tragedy, James Davidson was probably the most successful mariner-entrepreneur to navigate the industrialization of the Great Lakes. Although many Davidson-built ships ultimately wrecked, including seven in Wisconsin, he never lost a ship under his command and experienced only four complete losses in more than fifty years as a managing ship owner.

In the post–Civil War period, Great Lakes builders responded to the opportunities in heavy industry and to the enlargement of canals and the dredging of deeper harbors by constructing new types of ships. Powered by steam or towed by steam vessels, the Great Lakes ships of this era demonstrate a striking originality of design that blended contemporary developments in marine engineering with commercial requirements and natural resources of the Great Lakes region. This period saw the introduction of the classic Great Lakes bulk carrier, the technological peak of wooden shipbuilding and its decline in the face of resource exhaustion, and the appearance of the larger and more efficient steel ships. The stories and archaeological remains associated with the tragic schooner *Lucerne*, James Davidson's giant wooden bulk carriers such as the *Australasia*, and their contemporaries all reveal intense pressures of change and of new opportunities as the country placed heavier demands on the natural resources and productive capacities of the upper Midwest farms, mills, and factories.

Ironically, industrialization made more people dependent on Great Lakes ships than ever before—to carry crops, fuel, ore, building materials, manufactured goods, and other commodities. But for most, the new

The remains of the *Australasia* are located about two miles south of the north point of Whitefish Bay, off the Door Peninsula. TAMARA THOMSEN/WHI IMAGE ID 140131

dependence was indirect, and as the twentieth century took shape, the contributions of Great Lakes ships and of maritime people became lost in the loud cultural noise and economic rationalization of industrial America. By the 1930s, the living memories of the wooden age and of the Atlantic maritime culture were declining into nostalgia and folktale, and the contributions of the Great Lakes and its maritime people to the full dimensions of Wisconsin, midwestern, and US national history were diminished or forgotten. Alluring, exciting, and fragile, Wisconsin's historic shipwrecks, when properly studied, interpreted, and responsibly experienced by the public, act as a powerful antidote to the historical amnesia and cultural insularity that has undervalued the midwestern experience in North American and Atlantic history. This book is a maritime history inspired by Wisconsin shipwrecks.

Chapter One

THE GREAT LAKES AS AN ATLANTIC MARITIME FRONTIER

In 2005, Tamara Thomsen, a professional underwater photographer and historic preservation specialist with the Wisconsin Historical Society, took some exceptional photographs of the shipwrecked schooner *Home*. Good shipwreck photographs are rare, and Thomsen is a genuine artist. One of Thomsen's photographs casts the ship in an especially ghostly yet majestic aspect. Showing bow stem and forward port side, the image evokes a time and place that for most people does not suggest Wisconsin or the Midwest. Upright and resting against the side of the hull is a beautiful anchor so well preserved as to look more like a prop from a pirate movie than a feature of an archaeological site in the midwestern United States. Launched in 1843, the *Home* visibly embodies aesthetic principles and systems of technology brought to the Great Lakes by peoples from maritime communities on both sides of the North Atlantic. The fundamental Atlantic features on display in the wreck of the *Home* are also discernable on every historic shipwreck built of wood on the Great Lakes.

During much of the nineteenth century, the maritime technologies employed on the Great Lakes differed little in essence from those of the other coastal regions of the United States and North Atlantic. What was different, however, was the distance and, in the beginning, the physical isolation of the Great Lakes from the Atlantic Ocean. To exploit fully the Great Lakes, Europeans and Euro-Americans gradually transformed them from a group of isolated bodies of fresh water in the center of the North American continent into a westward extension of the Atlantic Ocean. More than

The ghostly bow of the schooner *Home* is a clear reminder of the Great Lakes Atlantic Ocean heritage. TAMARA THOMSEN/WHI IMAGE ID 119678

a thousand miles from the nearest drop of ocean water, they re-created the ships, ports, and systems of business, labor, and defense that characterized the Atlantic maritime world. In deliberate and measured steps, they redefined the geography to eliminate physical separation between the individual lakes, replacing them with a single inland sea—one connected physically and through ties of nation, culture, and economy to the Atlantic World.

Over the past three decades, Atlantic history has emerged as one of the most influential new paradigms shaping the work of American and European historians, and the idea of an "Atlantic World" is one of its most important concepts. Atlantic history and the Atlantic World emphasize cultural, economic, political, and ecological connections that transcend national boundaries. Ships and maritime people played a central role in the creation, transmission, and maintenance of these long-distance relationships that defined the Atlantic World. This book emphasizes the contributions these Atlantic maritime people and the Atlantic maritime culture brought to the Great Lakes. The term *Atlantic maritime culture*, as used in this book, includes technologies, skills, business practices, legal forms, language, and social outlooks that connected the seafaring peoples of the North Atlantic and ultimately those in the Great Lakes as well. The histories emerging from the study of Wisconsin shipwrecks, including this book, illustrate the direct and long-term influence of Atlantic maritime culture on the Great Lakes during the nineteenth century.[1]

A diver investigates the wreckage of the *Home*, which wrecked in 1858 when it collided with the schooner *William Fiske* in a dense fog southeast of Manito-WOC. TAMARA THOMSEN/ WHI IMAGE ID 119831

Investigating shipwrecks reveals the strength of these maritime connections and the extent to which they penetrated the popular consciousness of midwesterners during the frontier period. Wisconsin's early maritime iconography and the history of specific shipwrecks suggest that during that frontier period, people understood the Great Lakes in oceanic rather than inland terms. In describing the lakes, lake craft, and mariners, knowledgeable people applied the language of the sea. Through common social networks and labor pools, by adapting old and emerging maritime technologies, and by redefining the geography through canals and steamboats, the maritime people who came west deliberately drew the lakes and ocean together into a common maritime network whose oceanic foundations were more visible to them than to their twentieth-century descendants.

Traditional maritime historians often identify the years between 1815 and 1860 as the peak of American maritime enterprise.[2] During these same years, however, millions of men, women, and children left the Atlantic states and poured into the Old Northwest and Mississippi valley frontiers. There they laid down the foundations for the midwestern United States. The shipwrecks discussed in subsequent chapters illustrate the intimate connections between American enterprise on the ocean and continental expansion in the Midwest. Observed through the lens of Great Lakes shipwrecks and the histories that they reveal, it becomes clear that navigable water shaped Euro-American settlement of the Midwest during the nineteenth century every bit as much as it had in the New England and Middle Atlantic colonies during the sixteenth and seventeenth centuries.

For these reasons, to understand fully the history of Wisconsin and the upper Midwest, one must turn to the sea—not just to the *Inland Seas,* a term used to describe the five Great Lakes, but also to the global sea discovered by the Europeans in the fifteenth and sixteenth centuries that connected all of the continents of the world. The Inland Sea that inspired the construction of thousands of ships and then claimed so many as wrecks is a different place from the pristine lakes carved out by the glaciers so long ago.

The Great Lakes across Millenia

Although small compared with the ocean, the Great Lakes are geographically impressive. Encompassing ninety-four thousand square miles of surface area and ten thousand miles of coastline, they contain about 18 percent of the world's fresh water. Today they constitute a single waterway, which with its St. Lawrence Seaway connections extends nearly two thousand miles to the Gulf of St. Lawrence and the Atlantic Ocean and reaches the province of Ontario and the states of New York, Pennsylvania, Ohio, Michigan, Indiana, Illinois, Wisconsin, and Minnesota.[3]

Created by glacial scour, the Great Lakes are geologically young and took their present form about twelve thousand years ago when the warming earth melted the ice and uncovered deep basins that filled with water. People have inhabited the shoreline for at least seven thousand years, possibly longer, using the lakes as a source of food and means of transportation. Direct evidence of lake crossings can be traced to as early as 5000 BCE and to the copper transported from mines at Isle Royale in western Lake Superior by these early inhabitants of the Great Lakes region.

After thousands of years of human use, European explorers brought rapid change to the lakes. In a remarkably brief period, European mariners, driven by their quest to find sea routes to Asia, redefined the world map. Breaking out of well-trodden zones in the eastern North Atlantic and Mediterranean basin, they sailed south and east around the Horn of Africa and into the Indian Ocean. In 1492, Christopher Columbus became the first of many explorers to look for a westward route to Asia. Between 1519 and 1522, a Portuguese expedition organized by Ferdinand Magellan found a water passage between the Atlantic and the Pacific. This first circumnavigation of the globe proved the vastness of the Pacific and revealed that all of the earth's oceans were in fact one connected body of water.[4]

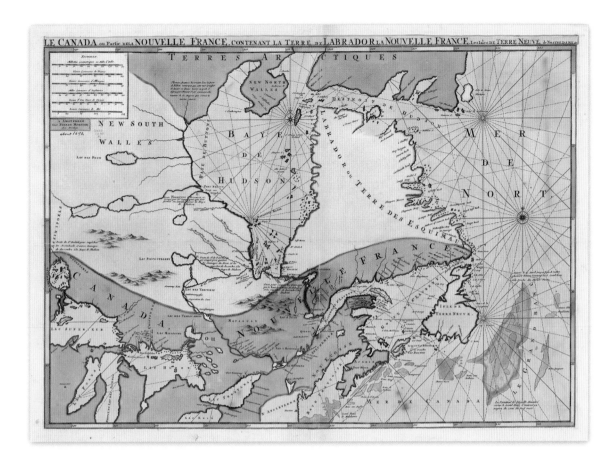

LE CANADA ou Partie DE LA NOUVELLE FRANCE, CONTENANT LA TERRE DE LABRADOR LA NOUVELLE FRANCE, Les Isles DE TERRE NEUVE, &c NOSTRE DAME &c.

Magellan's route into the Pacific was dangerous, long, and difficult. European explorers continued to search for a shorter and more direct water route, one that cut across the northern part of the American continent. This goal first led French explorers up the St. Lawrence River and into the interior of North America. In 1535, Jacques Cartier led an expedition that reached the site of present-day Montreal. During the journey, he heard reports of a large freshwater body farther to the west. This intelligence found its way into the work of influential exploration geographers, including Gerard Mercator and Richard Hakluyt, who believed this fresh water might form part of a larger Northwest Passage to the Pacific Ocean. Some seventy years later, in 1615, Samuel de Champlain finally reached Lake Huron. The process of incorporating the Great Lakes into the world ocean network and into Western Europe's economic, political, and cultural web had begun.

The seductive idea of a water passage to the Pacific continued to attract Europeans to the region, but the rich natural resources of fur, fish, minerals, and forests, combined with endless miles of navigable rivers and lakes,

This unusual map by Pierre Mortier combines the features of land and sea cartography and clearly situates the Great Lakes within the Atlantic maritime world of 1696. WHI IMAGE ID 123388 convinced many to stay. Although the French continued to search for a Northwest Passage to the Pacific well into the 1700s, they also explored the Great Lakes and the Mississippi valley, inspired by the increasingly lucrative fur trade. By the mid-1600s, French fur traders, Catholic missionaries, and allied indigenous people formed a cultural and economic network that stretched across a new maritime frontier.

A symbolic turning point in the oceanic transformation of the Great Lakes came in 1679 when René-Robert Cavalier, sieur de La Salle, constructed the *Griffon*, the first true Atlantic-style ship on the Great Lakes. At less than 50 feet in length, the *Griffon* was not a large vessel but carried five guns and a vast cargo compared with canoes.[5] During the seventeenth century, imperial power depended on gunned ships, and La Salle intended to use the *Griffon* to help consolidate French military and economic control over the Great Lakes and Mississippi valley. On her maiden voyage, the *Griffon* sailed to an outpost near the entrance of Green Bay, believed to be what is now Rock Island, Wisconsin, and discharged supplies and shipbuilding tools. On September 18, 1679, the ship sailed into the teeth of a violent storm and wrecked, possibly becoming the oldest shipwreck of Atlantic design in Wisconsin waters. The *Griffon*'s final resting place remains a mystery. Over the years, legions of shipwreck hunters have reported finding the wreck at different locations across the Great Lakes.

During the two hundred years after Champlain's arrival at Lake Huron, the Great Lakes region became a dangerous place where life was marred by international and intercultural violence. Rampant western diseases ravaged the Great Lakes Indians, as did incessant warfare. On the international front, the region became a conflict zone between empires—European and indigenous. Of the many imperial wars fought between European maritime powers from 1650 to 1815, three major ones had significant effects for the Great Lakes. In the Seven Years War, also known as the French and Indian War (1757–1763), England wrested control of Canada and the Great Lakes from France. Although retaining control of the Great Lakes throughout the Revolutionary War (1775–1783), England ceded the southern half of the lakes to the United States as part of the peace terms. The English, however, continued to occupy the southern lakes until 1796. Even after leaving, they enlisted their Indian allies to prevent widespread settlement by US citizens.

In dramatic fashion, the War of 1812 (1812–1815) ended the two centuries of international and intertribal conflict on the Great Lakes. Sparked by conflicts over economic sanctions that crippled US trade and unlawful impressments of American sailors into the British Navy, the war brought

the Great Lakes more fully into the consciousness of American politicians and military leaders. Command of the western lakes, as the Great Lakes plus Lake Champlain were known during this period, meant controlling the best invasion routes between the United States and Canada.

Both sides had Indian allies and access to hundreds of large Indian canoes and other smaller watercraft capable of carrying supplies and an invasion force. American and British military leaders realized, however, that military control of the Great Lakes would depend not on canoes or bateaux but on traditional naval power. At the beginning of the war, the British had four armed vessels in service, the largest mounting eighteen guns. The Americans, by contrast, had but one fourteen-gun brig, primarily used to carry supplies.

The naval arms race that followed transformed the Great Lakes from a primitive frontier on the periphery of the Atlantic maritime world to the center of innovative shipbuilding and enterprise. Noah Brown and Henry Eckford, the men charged with constructing the American fleets, were among the country's shipbuilding elite, and they brought advanced skills, a tradition of innovation, and nearly eight hundred maritime artisans—ship carpenters, caulkers, iron workers, sail makers, and more—to the shores of the eastern Great Lakes and Lake Champlain. On both sides of the border, the war brought a critical mass of Atlantic-based maritime talent to the Great Lakes.

Historians of technology have identified social networks as critical for understanding the process of innovation and the diffusion of new technologies. America's elite shipbuilders were part of a close-knit but highly competitive social and professional network that provided its members access to information and a forum for the exchange of ideas. The war extended this network into the Great Lakes, where it took firm root and exercised a powerful shaping force on Great Lakes shipbuilding throughout the wooden age.[6]

Henry Eckford was one of the most important American shipbuilders of his time. Eckford brought a large retinue of shipwrights and skilled artisans westward from New York City to build navy ships at Sackets Harbor on Lake Ontario. Among them, Isaac Webb, Stephen Smith, and Jacob Bell were or soon would be numbered among America's elite shipbuilders. Largely through Eckford's apprentices, between about 1820 and 1860, New York City reigned as the world's most important wooden shipbuilding center.[7] Eckford exercised a comparable influence on the Great Lakes, particularly in the construction of large steamers. Other Eckford men on the lakes included Henry Teabout and William Smith, who teamed up to build the Canadian steamer *Frontenac*, the first Canadian (British at the time) steamboat on the Great Lakes, completed in 1816.[8] A Mr. E. Merritt built the steamboats *Constitution* and *Henry Clay* in Black Rock (part of present-day Buffalo) in 1824 and 1825. Stephen Smith's apprentice, John Englis, built two steamers, the *Milwaukee* and the *Red Jacket*, at Buffalo in 1837 and returned in 1853

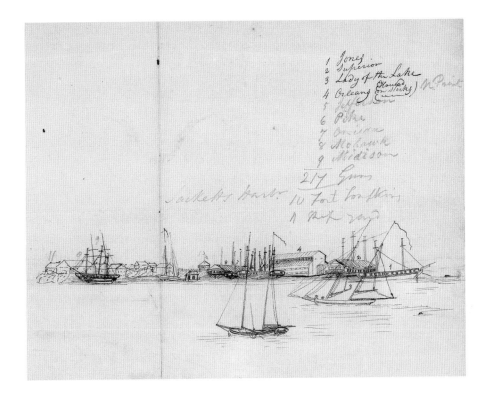

Sackets Harbor

1 Jones
2 Superior
3 Lady of the Lake
4 Orleans (on stocks) M. Point
5 Jefferson
6 Pike
7 Oneida
8 Mohawk
9 Madison

217 Guns
10 Fort Tompkins
A Ship yard

to build the enormous palace steamers *Western World* and *Plymouth Rock*.[9] A later Eckford apprentice, Jacob Banta, became partner with Benjamin Bidwell at Buffalo's most important shipyard, Bidwell & Banta. Beginning in the 1830s, Banta designed many distinguished steamboats, among them the *Niagara*, a palace steamer that burned in Wisconsin in 1856 and the subject of the next three chapters. In the 1850s, Bidwell & Banta hired an orphaned teenager named James Davidson, whose giant wooden bulk carriers would mark the conclusion of the wooden ship era on the Great Lakes at the beginning of the twentieth century.

Noah Brown's influence on the history of Great Lakes shipbuilding was more direct, if less pervasive. He and his brother Adam were former carpenters turned shipbuilders and got in on the ground floor of early steamboat construction. Returning from the lakes to their New York shipyard during the war, the Browns built the hull for Robert Fulton's *Demologos*, the US Navy's first steam warship, launched in October 1814.[10] After the war and Fulton's death in February 1815, the Browns became the principal builders of the Fulton Steamboat Company's vessels. In this capacity, they likely had a hand in building the first American steamboat on Lake Ontario, completed April 1817. The steamer *Ontario* was built using government wood at the navy shipyard. The design was based on the steamboat *Sea Horse*, the first built for service on rougher coastal ocean waters.[11] In 1818, Brown went west again and built the famous *Walk-in-the-Water*, the first American steamboat on the upper Great Lakes ("upper" meaning above Niagara Falls—Lakes Erie, Huron, Michigan, and Superior). When that vessel wrecked in 1821, Brown constructed its replacement, the steamboat *Superior*. Brown not only built the steamers but also held partial ownership and took an active hand in their management. In terms of shipbuilding lineage, the previously mentioned Benjamin Bidwell worked under Brown during the construction of one or more of these early steamboats.[12]

The War of 1812 attracted a unique group of shipbuilders. The Browns, along with Eckford and his many apprentices, combined a pragmatic approach to shipbuilding with exceptional creativity and resourcefulness. The context of war on the freshwater frontier allowed such characteristics to flourish as these builders had the room and impetus to innovate in shipbuilding methods and design that they would never have had in the civilian shipyards of New York or Boston. This spirit of innovation sparked by the war became a lasting part of shipbuilding culture on the Great Lakes.[13]

While the New York shipbuilders built the pioneer steamboats on both the Canadian and American sides of the lakes, they were not the only Atlantic transplants to bring new technologies and skills to the maritime frontier after the War of 1812. Peace brought an influx of maritime New Englanders, including many from Connecticut, such as Captain Augustus Jones Jr., who

This postcard celebrates the one-hundreth anniversary of Robert Fulton's *Clermont*, widely regarded as the first commercially successful steamboat. JIM WARD COLLECTION, SOUTHERN LANCASTER COUNTY HISTORICAL SOCIETY

Built in 1818, the *Walk-in-the-Water* became the first steamboat to navigate Lakes Erie, Huron, and Michigan. This drawing was penned by Samuel Ward Stanton in 1895. Stanton died on the *Titanic* in 1912. WHI IMAGE ID 139123

migrated to northern Ohio. Probably no other family had more influence on Great Lakes shipbuilding during the nineteenth century than the Jones family, and certainly none left a larger imprint on the marine archaeological record. Of the nearly five hundred Wisconsin Great Lakes shipwrecks for which we can identify the builder, Augustus Jones or one of his five sons built thirty-five. Shipwrights related to the Joneses, or who learned their trade under one of the Jones family members, built many others.

Augustus Jones was born in Essex, a part of the Old Saybrook Colony of Middlesex County, Connecticut, in 1781. By the time of his birth, Essex was an important center of shipbuilding and maritime commerce. Jones is an elusive figure. He built many ships and clearly traveled widely as a mariner before heading west, but he left a scant imprint in the historical record. He undoubtedly learned the shipbuilding trade in one of the village's well-respected yards. Between the American Revolution and the Civil War, Essex yards produced more than five hundred wooden ships. Related to some of the oldest and most prominent families in the area, Jones apparently prospered as a young man, possibly owning his own shipyard and shares of one or more vessels before the War of 1812.

During the war, Essex's reputation for shipbuilding and its association with American privateers attracted the attention of the British Navy, and they attacked the town on April 18, 1814, destroying twenty-eight ships and a total of $160,000 in property. Available accounts seem to exaggerate Jones's maritime holdings at the time of the raid, but he definitely held part ownership in at least one of the vessels destroyed, and the raid clearly devastated him financially. Lake lore states that Jones received a grant of government land near the mouth of the Black River in Connecticut Western Reserve of Ohio in compensation for his losses. No evidence of this grant has surfaced; however, many people from coastal Connecticut headed to the Great Lakes region after the war, where they joined others who had already established themselves in commerce and shipbuilding. When Jones migrated west, he joined a small but well-established network of shipbuilders, mariners, and businessmen with roots in Atlantic culture.

Augustus Jones and his brother-in-law Enoch Murdock relocated to Black River, Ohio, in 1818 where they teamed with another Connecticut-born shipbuilder, Fairbanks Church, to build the sloop *General Huntington*. Augustus Walker, a captain, shipbuilder, and entrepreneur whose career on the Great Lakes began in 1817, credited both Church and Jones with dramatically improving the design of Great Lakes commercial vessels. He goes into particular detail about Jones's schooners. Broader of beam but of less draft than comparably sized craft of other builders, the Jones schooners proved more seaworthy and could carry more freight on shallower water. Jones especially excelled in sparring and rigging. The wider distance between the masts and the distinct cut of sails provided the enhanced maneuverability so crucial for working off the dangerous lee shores common on the lakes.[14] Seaworthiness, maneuverability, and the ability to carry cargo over shallower water represented, in the long run, greater profits and enhanced safety.

MARRIAGE OF THE WATERS

One of Augustus Jones's early vessels, the schooner *Eclipse* built in Vermillion, Ohio, in 1823, was lost near Green Bay in 1843. Although the wreck has not been located, a contemporary description of its launching printed in the *Sandusky Clarion* supports Captain Walker's later assessment of Jones's work:

> *ECLIPSE* is 63 feet long, 18 feet 6 and a half inches beam, and 58 tons burthen, and is fitted and painted in a style highly creditable to the ingenuity of the builder and the enterprising owners. The cabin and stateroom are neat, airy and convenient. In short we are highly gratified with this specimen of improved naval architecture on Lake Erie.[15]

Appearing below this description, two additional articles underscore the exciting context of maritime expansion on the Great Lakes in 1823. The same issue of the newspaper that described the schooner *Eclipse* also contained an article reporting on the progress of harbor work for the much anticipated Erie Canal, then just two years away from completion.[16] The opening of the Erie Canal in 1825 is the most important event in modern Great Lakes history and a cardinal moment in the history of the United States.

The Erie Canal established the first navigable water connection between the Great Lakes and New York City. In physical terms, the canal seems more like a garden hose than the "artificial river" it was billed as in the 1820s. Just four feet deep in its original incarnation, with locks ninety feet long and fifteen feet wide, the canal spanned mountains, marshes, rivers, and streams on its 363-mile track between Lake Erie and Albany, the head of deep-water navigation on the Hudson River. In so doing, the canal remade the West and helped transform New York City into America's greatest metropolis. Even before the final segments of the canal opened in 1825, tolls collected across New York exceeded $1 million, and there was little doubt that the Erie Canal would prove a phenomenal success. Anticipation of the canal brought more settlers, including shipbuilders such as Augustus Jones, and encouraged the construction of both sail and steam vessels on the Great Lakes.

The ceremonies and celebrations associated with the completion of the Erie Canal in 1825 underscore the maritime nature of the enterprise and the deliberateness of the builders in connecting the Great Lakes with the

This view of boats passing through Lockport, New York, reveals the small original dimensions of the Erie Canal. Little more that a ditch, the Erie Canal opened up the Great Lakes and changed the course of American history.
WHI IMAGE ID 101359

larger maritime world. During the "marriage of the waters," an ornate boat carried a keg of Lake Erie water down the canal to New York City, where it joined other kegs taken from many of the world's other great rivers. Following high-toned speeches, officials solemnly poured these kegs into the ocean, joining them both physically and metaphorically. A more direct statement of the maritime cultural underpinnings of this union are the commemorative medallions that depict the sea god Neptune with his arm draped over the shoulder of Pan, whose goatlike anatomy made him a fitting representative of the fertile lands surrounding the inland seas.

This gold commemorative medallion showing the gods Neptune and Pan was created to celebrate the uniting of the Great Lakes with the Atlantic Ocean by the Erie Canal in 1825. HERITAGE AUCTIONS, HA.COM

Even before the entire length of the Erie Canal opened, its fame and early success within New York spawned a canal-building craze. In 1825, Ohio began a massive canal-building program. The Ohio and Erie Canal linked the Ohio River to Lake Erie at Cleveland in 1833 and set the stage for that city's rise to regional prominence. In 1845, another state canal connected Cincinnati to the Lake Erie port of Toledo. In 1836, Illinois began constructing the Illinois and Michigan Canal. When finally completed in 1848, the canal connected the Mississippi River (via the Illinois River) to Lake Michigan at a swampy settlement called Chicago. Although many of the canals built during the canal-building craze failed to show a direct profit, these artificial waterways unlocked the economic capacity of the Great Lakes region and helped solidify its urban footprint. Buffalo, Cleveland, and Chicago gained much of their regional dominance through the early canals.[17] Milwaukee, although not a canal city, also owed its early commercial success to the canals.

The construction of deep-water canals, designed to accommodate lake vessels, also proved critical. The Welland Canal, completed in 1829, allowed vessels to bypass Niagara Falls and transit between Lake Ontario and Lake Erie and into Lakes Huron and Michigan. As early as 1832, the completion of the Rideau Canal allowed modest-sized craft to transit between Lake Ontario and the St. Lawrence River and thence to the Atlantic Ocean. In 1855, a canal at Sault Ste. Marie completed the Great Lakes network by allowing ships to pass into Lake Superior. In time, the Sault Locks (also known as the Soo Locks) would become the most important industrial canal in the world. Canals transformed the Great Lakes from a series of enormous but isolated lakes into the "American Mediterranean," a large sea surrounded by land but connected to the Atlantic Ocean.

In an 1873 engraved image of a lock and dam at Sault Ste. Marie, two men are shown operating the sluice as ships wait on the opposite side of the lock. Another man is waiting with a rope with which to tie off an entering vessel. WHI IMAGE ID 96402

WISCONSIN AS AN ATLANTIC MARITIME FRONTIER

In 1836, the US Congress created the Wisconsin Territory out of a large section of frontier that extended from the western coast of Lake Michigan to the banks of the Missouri River. Despite its vast size, the new territory contained fewer than 12,000 non-Indian people. Growth, however, proved rapid. By 1842, the population reached 44,478, and three years later it had climbed to 155,277. In 1847, the year before statehood, Wisconsin added more than 50,000 persons. In 1850, the first federal census of the new state of Wisconsin counted a white population of 305,390 persons.[18] A large percentage of these settlers traveled west via the Erie Canal and the Great Lakes.

For the early white settlers, the significance of the water in exploiting the land was obvious, and its influences are visible in the design of Wisconsin's civic iconography. An 1837 etching of a proposed territorial seal for Wisconsin found in the visual archives of the Wisconsin Historical Society provides a potent illustration of the links between land, water, and human ingenuity. In the foreground, the seal features a highly accurate engraving of a late-model plow overlaid by lush grain; in the background and center, an equally well-drawn sidewheel steamboat, its stacks furiously billowing smoke, charges south down the Lake Michigan coast.

The first official territorial seal, created by William Wagner and adopted in 1838, offered a more crowded and representational version of the land and water connection in frontier Wisconsin. In addition to symbols of mining

In this proposed 1837 Wisconsin territorial seal, a Great Lakes palace steamer and steel plow symbolized civilization and technological progress. WHI IMAGE ID 141204

and agriculture, the seal featured Lakes Michigan and Superior, a sailing vessel, a steamboat, a lighthouse, and an anchor.[19] In 1851, Wisconsin adopted a new state seal, one that in most respects mirrors the seal now emblazoned on the state flag. The new seal featured two human figures, a yeoman resting his hand on a pick, and a mariner holding a coil of line. They personify the land and water sectors of Wisconsin labor. The specific representation of the mariner is important. Nothing marks him as a frontier or freshwater sailor. To the contrary, the tarpaulin hat, jacket, trousers, and scarf identify him as an Atlantic sailor, part of an international culture of deep-sea mariners found throughout the world during the mid-nineteenth century. Sailor-authors of the day, such as Herman Melville and Richard Henry Dana, provided their readers with colorful, accurate, and intimate descriptions of this group, which had long held a distinct place in the popular imagination and culture.[20] Selecting this highly specific version of a sailor again suggests the vibrant maritime cultural connections and overlapping labor pools between the Atlantic Ocean and the Great Lakes.

The sailor with his anchor, coil of line, tarpaulin hat, and distinctive jacket, trousers, and scarf brought an Atlantic cultural flavor to the 1851 Wisconsin state seal. WHI IMAGE ID 141205

The stories revealed through studying shipwrecks demonstrate that during the frontier period, Wisconsin depended on the Great Lakes in clearly definable and central ways. A high percentage of the population had direct experience with the lakes, whether as a passenger, a merchant, a city builder, or a maritime laborer or entrepreneur. The railroad, a competing and complementary form of transportation, had yet to make a significant impact on the state in 1851. Atlantic maritime people built and operated the ships that became physical links connecting Wisconsin with the Great Lakes region, the Eastern Seaboard, and the larger Atlantic World. In similar ways, Wisconsin's more than six hundred historic shipwrecks preserved by the waters of Lakes Michigan and Superior provide tangible evidence of the Great Lakes frontier's Atlantic maritime foundations.

Chapter Two

CHARLES REED AND THE EARLY YEARS OF GREAT LAKES STEAM

On the afternoon of July 7, 1848, Thomas A. Morris, a respected bishop in the Methodist Episcopal Church and a passenger on the palace steamer *Niagara*, stepped ashore in Milwaukee, the largest city in the new state of Wisconsin. A keen-eyed observer prone to quick and sometimes harsh judgments, Morris offered a guarded approval of Milwaukee, noting the "general air of cheerfulness" created by its cream brick and white-painted houses. A place he deemed respectable, the young community's many improvements testified to its future as a thriving commercial city. The cheerfulness and optimism of Milwaukee contrasted sharply with the ominous cultural relics Morris had observed earlier that morning while steaming down the Wisconsin shoreline of Lake Michigan. At Sheboygan, he noted the "scanty remains of the steamboat *Phoenix*," a craft that had burned in November of the previous year, killing more than two hundred people. Nearing Milwaukee, he surveyed the broken timbers of the steamer *Boston*, driven ashore by a powerful nighttime storm in 1846. Morris somberly noted, "These manifest tokens of destruction strongly admonish the passer-by of the perils of the lake."[1]

Bishop Morris made his observations from the deck of the *Niagara*, a sidewheel palace steamer and one of the largest and best-appointed vessels then in service on the Great Lakes. When placed into service in 1846, the *Niagara* was one of only two vessels on the Great Lakes to measure more than 1,000 tons in burden and the world's second-largest freshwater steamer. Although later eclipsed by even larger palace steamers, the *Niagara*

The propeller steamer *Phoenix* burned to the waterline a few miles north of Sheboygan, Wisconsin, when boilers overheated and set the engine room ablaze on November 11, 1847. The ship carried just two twenty-person lifeboats, and more than two hundred perished in the cold water, most of them Dutch immigrants traveling to homes in Wisconsin. WHI IMAGE ID 139124

remained a popular and important vessel until its tragic destruction by fire a few miles north of Port Washington, Wisconsin, in 1856.

Today the *Niagara*'s charred and scattered wreckage rests on the bottom of Lake Michigan, fifty feet below the surface—another token of destruction from the era of frontier steam on the Great Lakes. Despite infestation with invasive mussels and the picking over by early generations of scuba divers, the *Niagara* retains impressive archaeological integrity. Powerful, skillfully shaped timbers, a dramatic walking beam engine, and massive boilers speak to the *Niagara*'s contemporary reputation as one of the finest steamers on the Great Lakes. Built during the middle years of pioneer steam on the Great Lakes, the ship possesses architectural and engineering characteristics associated with the early steamers of the 1830s and the largest palace steamers of the 1850s. A midpoint vessel in the evolution of Great Lakes sidewheel steamers, the *Niagara* allows us to look backward and forward in time with equal facility.

During the 1990s, archaeologists from the Wisconsin Historical Society, assisted by many volunteers and with support from the University of Wisconsin Sea Grant Institute, documented the archaeological remains of the *Niagara*. That work, in combination with subsequent historical research, resulted in the wreck's listing on the National Register of Historic Places in four categories at the national level of significance. As official historic designations go, short of being deemed a National Historic Landmark (a complex

Encrusted with invasive mussels, the remains of one of the paddlewheel assemblies and the diamond-shaped walking beam are well preserved amongst the wreckage of the palace steamer *Niagara*. TAMARA THOMSEN/WHI IMAGE ID 140395

and often political process), the wreck of the *Niagara* received the highest marks possible.

Every historic shipwreck in Wisconsin can tell important stories, but only a few such as the *Niagara* provide so many different connections to an entire era in American history. Important in Wisconsin's past, the vessel reflects regional and national patterns in its association with westward expansion, and in the history of steam and transportation technology. At the broadest level, the *Niagara* embodies the deep connection between the frontier Great Lakes and national and international currents in American and Atlantic history.[2]

CHARLES REED AND THE EVOLUTION OF STEAM TECHNOLOGY

To set the stage for the construction of the *Niagara* and for the palace steamer era of the 1840s and 1850s, it helps to revisit the early decades of pioneer steam navigation on the Great Lakes. The merging of frontier opportunities and environmental conditions with the most imaginative minds in the world of wooden shipbuilding and steam engineering yielded extraordinary results. Between the 1816 launch of the *Frontenac* and the mid-1850s, an extraordinary class of sidewheel passenger steam vessels evolved on the Great Lakes. When built, they ranked among the largest, most elaborate, and fastest steamers in the world. The role that these vessels played in setting the cultural and economic footprint of the Great Lakes region is now little remembered but should not be undervalued, for in their own time their importance was well understood and celebrated.

During the pioneer period, steam provided the power that shaped the western frontier. James L. Barton of Buffalo, in a widely circulated 1847 essay, described the transformation of the northern and western Great Lakes region from wilderness to civilization as seeming to occur "suddenly, as if by magic." The rapid change, he argued, happened "with the powerful aid of steam, and the indomitable enterprise, industry, and perseverance of free people."[3] Barton was far from alone in recognizing the amazing effect of steam on the frontier. In 1848, Wisconsin's statehood year, prominent western booster James Hall declared in an often reprinted statement that the "application of steam power to the purpose of navigation, forms the brightest era in the history of this country." To steam navigation, Hall attributed the rapid expansion of the western population and "the almost miraculous development of our resources."[4] Although he was from Cincinnati, the most vibrant of the western river cities, Hall's words held true for all of the Midwest, including Wisconsin during its territorial and early statehood days.[5]

Perhaps no one man exercised more influence on the course of marine steam on the Great Lakes during this period than Charles Manning Reed (1803–1871), entrepreneur, politician, and onetime owner of the largest fleet of steamers on the Great Lakes. From the 1820s through the 1850s, Reed's career, activities, and ambitions overlapped with the evolution of marine steam on the Great Lakes. He bought or commissioned the construction of many important ships, among them the *Niagara*.

Reed's life began on the maritime frontier. His grandfather, Seth Reed, a Massachusetts physician, revolutionary war hero, and shrewd land speculator, established the family at Presque Isle near present-day Erie, Pennsylvania, in 1795. Seth Reed soon died, but his son Rufus S. Reed established a trading post and tavern at Erie in 1796 and enmeshed himself in the region's fur trade. He lubricated his enterprise by trading alcohol to the Indian tribes, a technically illegal practice. Family lore proudly recounts Rufus Reed's circumvention of the Pennsylvania law prohibiting the sale of liquor to Indians "by the gill, quart, or barrel" by providing it "by the yard" in long hollow pipes.[6] Whether or not he established his empire on the foundations of liquor, the aggressive Rufus Reed became one of the wealthiest men in the Great Lakes region before his death in 1846.

Charles Manning Reed, the only child of Rufus Reed, was born in Erie in 1803. From his boyhood vantage point at Presque Isle, the young Charles must have watched with excitement as gangs of men from the East Coast built powerful warships and celebrated when the Americans emerged victorious at the Battle of Lake Erie. His early education is unknown but clearly excellent. As a teen, he left Erie to attend Washington and Jefferson College in western Pennsylvania. From there, he went on to Philadelphia to study law under Horace Binney, one of the country's leading attorneys. Reed passed the bar in Philadelphia in 1824 and returned to Erie, where he went into business with his father. With an analytical approach, he proved to be quick at discerning, seizing, and shaping the new opportunities offered by the Great Lakes and the surrounding region. By all accounts, Reed possessed a powerful mind, one that kept him at least one step ahead of the rapidly changing midwestern economy. In a century and region characterized by periods of boom and bust, Reed never faltered.

Portrait of General Charles Manning Reed, transportation pioneer and "Napoleon of the Lakes" during the palace steamer era. COURTESY OF THE ERIE CLUB (GENERAL CHARLES MANNING REED MANSION)

Reed's return to Erie coincided with a new era on the Great Lakes frontier. The Erie Canal was nearly complete, and the old impediments to western settlement—Indians and the British—had been neutralized. Old enough to remember life before the War of 1812 and to marvel at the *Walk-in-the-Water* in 1818, Reed possessed youth, vision, intellect, and wealth: the perfect ingredients to prosper in the West. During the next three decades, Reed opened stores and invested heavily in sailing vessels, banks, canals, railroads, and the development of Erie's early industries. Despite his diversified interests and a brief but significant political career, the public at large associated Reed with his steamers. Reed was the largest individual actor in early Great Lakes steam navigation, and by midcentury, the public began regarding him as "the Napoleon of the Lakes." Identifying all of the steamers Reed owned is difficult because he conducted most of his business through partnerships. Nevertheless, whatever the legal definitions of ownership, if a particular steamer involved Charles Reed, the public recognized him as the man in charge.

Steam navigation on the American Great Lakes began slowly, with only two boats built between 1818 and 1824. In 1825 and 1826, however, the new Erie Canal inspired the commissioning of at least seven new steamers.[7] Among those involved in this early steam rush were the Reeds, who invested in the ninety-five-foot steamer *William Penn* built at Erie, Pennsylvania, in 1826. Three years later, in 1829, Charles Reed organized a group of Erie businessmen who purchased the unfinished 120-ton steamer *William Peacock*. They finished the vessel and placed it into service between Erie and Buffalo. The enterprise proved an overall success but was one that underscored the dark side of pioneer steam. In 1830, an engine pipe failed and injected a cloud of pressurized superheated steam into the cabin, killing fifteen people and scalding many others. It was the deadliest steamer accident to occur on the Great Lakes up to that time. The dead were buried, and the *William Peacock* soon was repaired and returned to service. Undeterred by the accident, Rufus and Charles Reed continued to invest aggressively and systematically in steam navigation, establishing a shipyard at Erie and a network of wharfs, wood stations, and warehouses at strategic points around the lakes.[8] Their largest investments, overseen by Charles, involved the construction and operation of sidewheel steamers; among them were some of the largest ships on the Great Lakes.

In October 1832, Charles Reed launched the 380-ton *Pennsylvania*. Reed powered the vessel with two steam engines, a feature found on some of the better Long Island Sound boats of the 1820s but one that never gained favor on the Great Lakes. The two engines suggest that Reed, among other things, wanted a fast boat. Three years after its construction, Reed's advertisements claimed that the *Pennsylvania* "is not surpassed in speed, by any

A manifest lists boxes of manufactured and finished goods loaded as cargo on the steamboat *Pennsylvania* in Buffalo, New York, for October 1837 voyage to the frontier port of Chicago, Illinois. US CUSTOMS, GREAT LAKES RECORDS, BURTON HISTORICAL COLLECTION, DETROIT PUBLIC LIBRARY boat on Lake Erie."[9] During the *Pennsylvania's* debut 1833 season, more than sixty thousand passengers boarded steamers based in Buffalo, the hub of Great Lakes steamer travel, and more than two-thirds of them were headed west.[10] That year, Reed also purchased a partial ownership in the large new steamer *New York*.

In late 1834, the shipwrights at Reed's Erie yard laid down the keel for the 500-ton *Thomas Jefferson*. One observer described the new steamer, then the largest on the lakes, as "a monster of the deep;" another called it "a floating castle."[11] Although far smaller than the fully developed palace steamers of the 1840s and 1850s, the *Jefferson* possessed characteristics associated with these later vessels: large size (for the time), structural strength, high speeds, and stylish, even luxurious, cabins for first-class passengers. In June 1835, Reed accompanied the vessel on its maiden voyage to Lake Michigan. The trip coincided with a major federal land sale at Chicago, and Reed filled the steamer with high-paying speculators. Demand for the passage proved so heavy that Reed's officers commandeered the new vessel's steerage cabins to provide sufficient accommodations for the many first-class passengers. Reed invested heavily at the Chicago sale, purchasing business lots and waterfront property where he later built a substantial wharf and store. With property and facilities in Erie, Buffalo, and Chicago, and with several fine vessels, Reed had positioned himself to take the lead in some of the most profitable steamboat routes on the Great Lakes.[12] Reed's investments soon paid off. Passenger travel reached record levels in 1836 and 1837, with the number of people traveling on Great Lakes steamers each year topping, according to one estimate, two hundred thousand. The *Thomas Jefferson* is said to have paid for itself in its first season.[13]

When launched from the Reed shipyard in Erie, Pennsylvania, in 1834, the *Thomas Jefferson* was described as "a floating castle" and the largest steamer on the Great Lakes. WHI IMAGE ID 139125

DAVID STEVENSON

David Stevenson, a civil engineer from Scotland on a tour through the United States, was among the 1837 travelers. Only twenty-two years old, Stevenson belonged to a family engineering dynasty that specialized in the construction of lighthouses, including several built along the rugged coast of Scotland. For three months, Stevenson toured the country making detailed observations about American engineering and the natural environment of North America. He compiled the notes of his tour into the book *Sketches of the Civil Engineering of North America*, published in late 1837. Stevenson's professional training and knowledge of coastal and marine engineering and general maritime operations inform the book and offer a perspective that helps to place the Great Lakes of the mid-1830s into larger national and international contexts.

Stevenson traveled throughout the United States and in every region found interesting things to observe and describe. The maritime region that may have surprised him the most, however, was the Great Lakes. He wrote that when the Great Lakes are "viewed in connection with the River and the Gulf of St. Lawrence, by which their surplus waters are discharged into the Atlantic Ocean, ideas of magnitude and wonder are excited in the mind which it is impossible to describe."[14] In describing the lakes, Stevenson urged his readers to abandon any preconceptions about bodies of fresh water. The Great Lakes, he explained, are "vast inland sheets of water" and "bear a much closer resemblance to the ocean than the sheltered bays and sounds in which the harbours of the eastern coast of North America are situated."[15] In observing Buffalo's one-third-mile-long breakwater and masonry-covered piers, Stevenson noted, "one might fancy oneself in some sea-port, forgetting altogether that he is on some freshwater lake." He judged these and other Great Lakes harbor works as better able to resist "the fury of the winds and waves" than the comparable wooden structures found in the ports of the northeastern states. Although possibly at first appearing "superfluous" in their heavy construction, the engineer ascribed their strength to the power of the lakes and described the extraordinary natural violence evident in the Buffalo harbor works as well as in the many large uprooted trees, flung long distances from shore by wind and wave.[16] Given his intimate familiarity with the famously violent coasts of Scotland and Ireland, Stevenson's observations of the Great Lakes natural environment carry serious weight.

By 1837, steam navigation in the United States had passed its infancy, and Stevenson devoted more than a chapter of his book to it as "one of the most interesting subjects connected with the history of North America."[17] As he did throughout the book, Stevenson embraced a comparative approach in his investigations and identified three principal steamboat regions. The oldest was the northeastern United States, which encompassed the Hudson River, Chesapeake Bay, and the Long Island Sound corridor between New York City and southern New England. The area of the most activity included the western rivers the Ohio and Mississippi and their tributaries, served by hundreds of steamboats. And, finally, the western lakes were home to between forty and fifty steamers.[18] Although the boats in each region shared a common origin in the work and vision of Robert Fulton, the thirty years since the first Fulton boat had brought much regional differentiation and, in Stevenson's view, not all of it for the better.

In the northeast United States, the birthplace of steam navigation, Stevenson found ingeniously designed and well-constructed vessels. The primary design goal embodied in these vessels was achieving maximum speeds. He attributed the push for speed to the region's large population and the strong demand for rapid passenger service between major cities. The high speeds that northeastern steamers achieved and their general quality reflected, he believed, the intense competition among the region's many fine builders.

Stevenson offered harsher words for the steamboats of the western rivers, where he concluded that economy seemed "the only object." He judged the workmanship poor, dangerous, and even misleading. The gaudy splendor found in the first-class cabin, Stevenson ventured, "serves in some measure to distract the attentions of its unthinking inmates from the dangers that lie below them." Worse than the boats were their officers, "whose recklessness of human life and property is equalled only by their ignorance and want of civilization."[19] Hastily built and powered by high-pressure steam engines, the western riverboats had been the scenes of many horrible accidents. Built for economy first and then speed, but not for strength, the typical western river American steamers, according to Stevenson, lacked sails or real mariners with the skills and judgment required to navigate them safely. While the large river steamers of the early 1820s had retained distinctly Atlantic aspects with their raised sterns and substantial bows, these features gradually disappeared in favor of distinctive river-influenced designs. The environmental factors that shaped rapid river steamers included shallow-water navigation and the absence of large waves. In combination with frequent accidents, these contributed to a quick divorce between Atlantic and western river steamboat design.

Stevenson highlighted some notable exceptions to his general criticisms of American steamboats. A few of the boats that ran between New York and New England possessed seagoing characteristics, as did the steamers found on the frontier lakes. Stevenson visited at an excellent time to observe the latest in Great Lakes steam. In February 1837, at least seventeen new steamboats were under construction, including one, the *James Madison*, for Charles Reed. Described by a Buffalo newspaper as a "monster among our lake craft," Reed's new 181-foot steamer entered service as the largest on the lakes.[20] The new vessel impressed David Stevenson. He described it as a "strongly built vessel of 700 tons burden, drawing about ten feet of water." The favorable impression of the *Madison* was heightened by Stevenson's recent examination of "the many slender fabrics, drawing between three and five feet of water" along coasts and rivers of the northeastern United States. Indeed, the Great Lakes vessels were not like the other American steamers. As a class, they evidenced "in a greater degree the character of sea-boats." As Stevenson observed about the Buffalo harbor, the lake steamers possessed a structural strength and embraced designs that, at first glance, seemed inconsistent with inland water navigation. As with the harbor works, Stevenson attributed their characteristics to the requirements imposed by the "severe storms and formidable waves" found on the western lakes. Furthermore, in direct contrast with nearly all the other American steamers, but in keeping with European practices, lake steamers possessed working sails, and they were navigated by "persons possessed of a knowledge of seamanship."[21]

What Stevenson observed in the *James Madison* was a new, well-finished steamer that appeared "quite sharp, fore and aft" compared with other lake vessels. In a rare departure for Reed, he powered the new craft with a high-pressure steam engine built in Pittsburgh. Lighter in weight, requiring less space, and more economical to build, high-pressure engines also had earned a reputation for contributing to the deadly boiler explosions plaguing the western rivers. During times of tight competition for travelers, a high-pressure engine could scare off potential passengers. After building the *James Madison*, Reed returned to the more expensive low-pressure engines.

The *James Madison*'s design came from a Connecticut mariner and shipbuilder, Captain Fairbanks Church, who had been among the maritime Yankees who migrated to northern Ohio after the War of 1812. Church, who along with Augustus Jones was credited with bringing significant improvements to the design of Great Lakes sailing vessels during the 1820s, had built a number of fine steamers in the early 1830s. In Church and Jones, one can

This Henry Lewis painting of the Mississippi River steamboat *Grand Turk* illustrates the differences between river and Great Lakes steamboat designs, as described by David Stevenson. Built in 1848, the *Grand Turk* burned at New Orleans in 1854. WHI IMAGE ID 53740

find possible explanations for Stevenson's observations about Great Lakes steamboats such as the *James Madison*. The men building the best early Great Lakes steamers, such as the *James Madison*, came from the sea, and they conceived the pioneer Great Lakes similarly to the way Stevenson conceived them: as more ocean than lake. Recognizing the power of the Great Lakes and the dangers posed by high winds, steep waves, ice, and ever-present lee shores, they built strong boats that could, they hoped, survive when steam failed, even in the midst of a storm. In general, the lakes offered more dangers than the Hudson River, Long Island Sound, or the Chesapeake Bay. Maritime statistics collected by governments and insurance companies support the notion that the best pioneer shipbuilders understood the environment well, for the lake region saw a disproportionate share of shipwrecks and other marine accidents throughout the nineteenth century.

Charles Reed appears to have analyzed the Great Lakes in more economic than environmental terms. As expressed in ships, however, the environmental and economic mentalities are nearly impossible to separate. Although he grew up on the shores of Lake Erie and possessed years of experience with ships and shipping, Reed did not come from a maritime family. The Reeds' business was extracting and generating wealth from the western frontier. As economic opportunities evolved and changed in the region, the tools Reed needed to exploit them changed as well. The largest business tools of the day involved transportation: roads, canals, steamers, and railroads—and Reed invested in them all. In contrast with these other transportation tools, however, ships could embody almost human qualities and take on a clear public identity, one associated with its owner, builder, and officers.

Ships, especially complex and expensive ones such as the *James Madison*, merged the economic and social ambitions of the owners with the imaginations and resourcefulness of the shipbuilders. The builders and their employees made thousands of operational and technical decisions, but many of the largest conceptual decisions, including a vessel's relative size, type of engine, and cabin arrangements, came from, or at least required the consent of, the owner. The confluence of economic ambitions, emerging frontier opportunities, abundant natural resources, and the imaginative marine engineering talent available on the Great Lakes during the 1830s resulted in the rapid evolution of the region's steamers. This evolution, however, can be followed only in outline and not in detail. The archaeological study of more steamers in the years to come should fill in these gaps.

During the 1830s, investors such as the Reeds poured millions of dollars into Great Lakes steamers. For many, it paid off as the region began attracting tens of thousands of settlers, speculators, and entrepreneurs to the states and territories adjacent to the Great Lakes. Although Cleveland, Detroit, Sandusky, Erie, and even tiny Milwaukee had steamers registered in their customs districts in the 1830s, because of the Erie Canal, Buffalo was the principal steam transportation hub on the Great Lakes. From Buffalo, boats radiated out across the lakes to points as far west as Green Bay and Chicago. Within a short time, railroads paralleling the Erie Canal expanded the transportation conduit from the Atlantic coast to Buffalo, and by the late 1830s, passenger traffic on lake steamers had grown large. It was not unusual to have one, two, or even three thousand transient people milling around Buffalo (a city of fewer than twenty thousand persons) on any given day during the traveling season. One day in 1838, five thousand people departed westward from Buffalo, which was fast becoming "the New York of the western regions."[22]

In the case of Reed, wealth begot more wealth and economic and political influence. In addition to steamboats, he and his father invested in a wide array of enterprises, including banks, canal companies, and pioneer railroads. With more wealth came additional status and public responsibility, and their names are associated with a host of philanthropic activities in Erie.[23] The only son of a rich man, Reed led an active social and public life.

He enjoyed military display and purchased a commission as a colonel and later as a general in the Pennsylvania state militia. Reed embraced the military title as part of his identity, and the newspapers often referred to him as "General Reed." Membership in Reed's infantry company, it was said, brought decided advantages to the socially ambitious in Erie society. In the midst of building up the family enterprises, Reed found time to serve a term in the Pennsylvania legislature in 1837 and 1838 and, during the second year of his term, to marry Harriet W. Gibson of Watertown, New York.[24]

In 1839, Great Lakes steamers carried more passengers and cargo than ever before. At least eight steamers ran regularly between Buffalo and Chicago, with first-class boats making two scheduled round trips each month during the season.[25] Reed established himself at the top of this trade by operating four grand steamers, the *Thomas Jefferson*, the *James Madison*, the *Buffalo*, and the *Erie*, and possibly other smaller vessels as well. Recently returned from the Pennsylvania legislature, Reed drew this prescient compliment from a Buffalo reporter: "Mr. Reed understands how to do business. A few months ago, he married a beautiful wife, and now look at his steamboats. This is real enterprise, and he will in a few years be the richest man on the lakes."[26]

For some, wealth would have to wait or would never come at all. During the late 1830s and early 1840s, the Great Lakes economy tanked as the result of a severe national depression. Business on the lakes dropped to low levels, property values plummeted, and many recently rich western entrepreneurs lost their fortunes, with no small number going bankrupt. Reed, by contrast, continued his rise, turning some of his attention from the stalled world of steam navigation back to the fertile world of politics. Elected to the US Congress as a Whig in 1842, he pushed hard for internal improvements on the Great Lakes and especially for the harbor at Erie. Narrowly defeated for reelection, Reed returned home in 1845 and resumed an active interest in business. Once again, his timing proved fortunate, as the lake trades were fast recovering and steam navigation was moving into a new phase. Burgeoning emigration and a booming economy would bring more wealth and encourage the construction of new, vastly larger steamers. It was the beginning of the age of the palace steamer, and Reed would move fast in seizing the new opportunities.[27]

Chapter Three

BUILDING THE PALACE STEAMER *NIAGARA*

As traffic on the Great Lakes increased during the 1830s and 1840s, ship owners such as Charles Reed discovered the economic advantages of larger steamers. Carrying more passengers and cargo, the larger craft required relatively less power to achieve the required speeds and to deliver cargo. This translated into a savings in fuel, which the palace steamers consumed in immense quantities. The larger size also brought additional economies of scale as the number of hands required to operate the machinery and navigate a sidewheel steamer did not increase proportionally with size.[1] Most of the additional labor required to operate larger palace steamers involved service positions, the need for which increased in proportion to the number of passengers carried.

Shipbuilders on the Great Lakes faced major challenges in designing the larger vessels. During the palace steamer era, the most important limiting factor was the minimum available depth of water through the principal navigation routes. While a steamer floated with hundreds of feet of water below the keel during most of a cross-lakes voyage, at the strategic St. Clair Flats between Lakes Erie and Huron, the depth often dwindled to a few inches or less. The Flats became the principal navigation bottleneck of the 1840s and 1850s. Part of the St. Clair River Delta, the Flats combined shallow water and dangerous rocks with shifting sediments, making permanent navigation channels unreliable. A new sandbar or a slight piloting error could lead to an almost instant grounding.

Daily, seasonal, and long-term variations in lake levels added to the difficulties. Over multiyear periods, the average water level on Lake St. Clair, as measured since the late 1890s, has varied by up to six feet. Over the course of a single year and following a predictable cycle, lake levels typically rise and fall twelve to eighteen inches, with water levels lowest during the winter and highest during the summer. Dramatic and abrupt short-term changes can also occur over a span of days, hours, or even minutes. Protracted heavy winds cause an imbalance of water levels, raising them at one end of a lake and decreasing them at the other. Rapid changes in barometric pressure, a frequent prelude to a storm, also cause or contribute to the abrupt changes. When conditions quickly return to normal, the built-up water comes pouring back until it overfills the other side of the lake. This lakewide oscillation, called a seiche, may continue for days and cause water levels to change by several feet over a few minutes. The short-term fluctuations of water levels made crossing the Flats in a deep vessel a tense experience, one that required a combination of skill, guesswork, and luck. The longer seasonal cycles, however, created more worry for ship owners, who feared catastrophic disruptions in navigation. The Flats remained a consistent problem in the 1870s.

The longer steamboats offered better speeds and economies, but only if they could regularly ply between the ports of Lake Erie and Lake Michigan. The longer palace steamers had another problem, one shared by the later Great Lakes bulk carrier: adequate hull strength. A long ship is intrinsically weaker than a shorter one of similar breadth and depth. An experiment with a retractable metal tape measure provides insight into the long ship dilemma. When extended two or three feet, the metal tape will remain straight and run parallel to the floor. Extend it farther with no additional support, and it will abruptly buckle and bend toward the floor. Although water provides structural support that air does not, this support is concentrated in the most buoyant sections of the hull. Because the bow and stern are less buoyant than the hull of a ship, long ships tend to droop on each end, a process known as *hogging*. A related problem involves the heavy weight of steam machinery making the hull bend in the middle, an occurrence known as *sagging*. The designers of large palace steamers faced both problems: hogging at the bow and stern and sagging amidships from the concentrated weight of the engines and boilers.

Builders of deep-water vessels, designed to transit the open ocean and visit only deep harbors, could compensate, to some degree, for the weakness of longer hulls by building them with deeper draughts. A deeper hull offered more room to build in the longitudinal support needed to reduce hogging and sagging and to resist the twisting stresses generated by heavy seas. Just as an 8-foot-long 4-inch by 4-inch pine board will resist bending and breaking better than a 2-inch by 2-inch board of equal length, so it was with wooden ships. The longer the ship, the more the potential for hull failure.

AMBITION OF THE *EMPIRE*

The designers of palace steamers understood the physics of water and compensated by developing new methods for strengthening large ships. During a twenty-year period beginning in the early 1830s, the sidewheel passenger steamers on the Great Lakes nearly tripled in length and grew sixfold in tonnage. This growth came in waves. The 130- to 150-foot class of steamers built in the early 1830s gave way to the 170- to 180-foot vessels typified by Reed's *James Madison* and so well described by David Stevenson in 1837. The next wave proved more monumental than incremental. In late 1843, Cleveland, Ohio, businessmen D. N. Barney and James Smith engaged shipbuilder George Washington Jones to design and build America's largest freshwater steamer. At about 263 feet in overall length, the *Empire* exceeded the largest existing lake steamers by 60 to 80 feet in length and 400 tons in burden.

A remarkable example of marine engineering, the *Empire* not only changed the scale of Great Lakes and American steamboats, it did so with a decidedly regional cultural stamp. Although born in Connecticut, the builder, George Washington Jones, grew up and learned his trade on the shores of Lake Erie in northern Ohio under his father, Augustus Jones. Whereas most large Great Lakes steamers of the period possessed low-pressure steam engines manufactured on the Atlantic coast, the *Empire* featured a high-pressure engine and enormous railroad-style boilers built in Cleveland by the Cuyahoga Steam Furnace Company.[2]

In 1844, the massive *Empire*, the pride of Cleveland, Ohio, would have drawn attention at any harbor in the world for its size. The steamer's cabin architecture and decor also garnered enthusiastic praise. The main cabin, described as being of "princely magnificence," centered on an ornate open salon about 230 feet in length. Around the perimeter of the salon, seventy-two staterooms set a new standard of comfort for western travel with wide beds, double mattresses, and warm covers—a far cry from "the warm board and tape covering which has heretofore been called bed and bedding for the traveller."[3] A genuine wonder for its time and a remarkable leap forward in size, the *Empire*'s nautical architecture included details that drew some

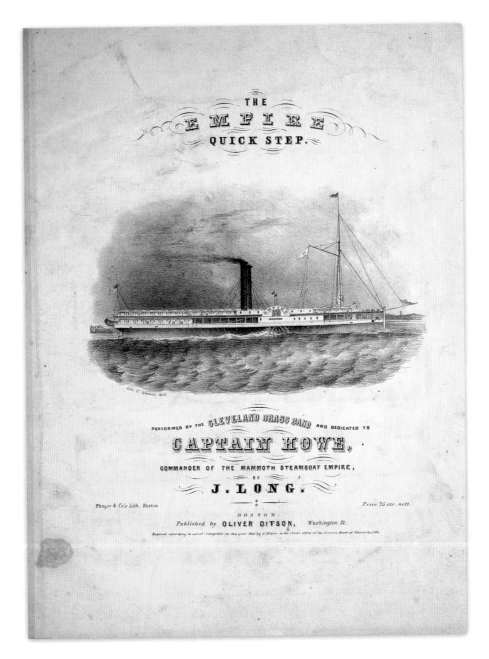

Composed to celebrate the launching of the *Empire* in 1844, "The Empire Quick Step" sheet music features a highly detailed illustration of the revolutionary steamer. WALTER LEWIS

note—a rarity in the coverage of steamboat launchings. George Washington Jones compensated for the potential weakness of the long hull and the heavy machinery by installing two semicircular arches that extended up and over the deck amidships.[4] Such arches, later known as *bishop arches*, became typical features on many Great Lakes passenger steamers. In addition, Jones covered the exterior and interior of the hull with 6-inch-thick planking.

The hull probably possessed other attributes that added to the *Empire*'s longitudinal strength and durability, but one thing is certain: Jones built a ship designed to absorb a great deal of physical abuse in addition to accommodating its great length and heavy machinery. The *Empire* was an economic and technological success, remaining in service for many years as a steamer and surviving as a tow barge into the 1870s.

The 1844 launching of the *Empire* raised the standard for first-class Great Lakes steamers in dramatic fashion. The following year, Charles Reed responded by commissioning the construction of two new large vessels, the *Louisiana* and the *Niagara*. This competitive maneuver was not merely a question of pride—although with Reed, public appearances clearly mattered. Business dictated his actions. Without large and luxurious new vessels, he could not maintain his commanding position in the growing trade between Buffalo and Chicago. Always analytical and thoroughly informed, Reed understood the regional economy and the technology required to seize current and future opportunities.

DESIGNING THE *NIAGARA*

The new steamers reveal a careful balance of imagination and pragmatism. At 245 feet in length, 33 feet in breadth, with a 14-foot depth of hold, and measuring nearly 1,100 tons, the *Niagara* rivaled the *Empire* in size, speed, and luxury. The *Louisiana*, although smaller at nearly 800 tons, was still a large ship by 1845 standards. Significantly, it came in at a bargain price of $55,000, a full $40,000 less than the *Niagara*. The savings came principally in the engine, which Reed recycled from his aging *Thomas Jefferson*. Equally dependable, but faster, and more seaworthy than the *Thomas Jefferson*, the *Louisiana*'s success embodied the recent improvements in the region's nautical architecture.[5]

Reed elected to build the steamers in Buffalo, New York, rather than at his own shipyard in Erie, Pennsylvania. Why he came to this decision is unclear; throughout his life, Reed worked tirelessly to build up Erie's economy. By 1845, however, the economic geography for large-scale shipbuilding had tilted far in favor of Buffalo. The city offered direct water access to New York City through the Erie Canal, a more developed manufacturing base, and a larger skilled-labor pool that included an extraordinary contingent of talented shipbuilders.

Reed contracted with the firm of Bidwell & Banta. By the 1840s, junior partner Jacob Banta enjoyed a growing reputation as one of the finest steamboat designers on the Great Lakes. Born in 1809 to a prominent ship-building family and trained in New York City by Henry Eckford, Banta migrated west and settled in Buffalo where he later joined Benjamin Bidwell as a partner in the oldest shipyard on the American Great Lakes. Operating as Bidwell & Banta, the yard launched dozens of sail and steam craft between the early 1840s and 1860, including several of the largest Great Lakes palace steamers. The firm's success with steamers came chiefly from Banta, who was remembered after his death as "a genius of high order" who possessed more "talent and taste in naval science" than his peers.[6] The firm's senior partner, Benjamin Bidwell also offered an impressive pedigree. Yet another Connecticut-born Great Lakes shipbuilder, Bidwell migrated before 1810 to the Buffalo/Black Rock area where he became the apprentice and later the partner of Asa Stanard. Stanard and Bidwell worked under Noah Brown in the building of the *Walk-in-the-Water* in 1818 and its successor *Superior*. In short, by 1845, Bidwell & Banta offered a combination of regional experience and innovation in steamboat construction that few could match.

Work on the *Niagara* began in the winter of 1845, with the massive hull taking nearly six months to complete, a rather long time for that era. On June 4, 1845, after multiple delays, a huge crowd gathered for the launch of the ship. Thousands of eyes watched with rapt attention as the long hull began sliding down the shipyard ways. About halfway through the launch, the ship struck a submerged timber and stuck fast. The abrupt stop caused the hull to bulge out amidships. The press noted that an "ordinary built boat would have broken . . . in two."[7] When considering palace steamers, the term *ordinary* has little meaning, but the incident did reveal the exceptional durability that would mark the ship's career.

For success, the *Niagara* required more than durability; it needed to offer speed, dependability, a high standard of luxury, and a hull capable of passing over the St. Clair Flats. Unlike descriptions of the pioneering *Empire*, contemporary descriptions of the *Niagara* offer only sparse details regarding its nautical architecture. The wreckage of the *Niagara*, however, provides a wealth of insights into its design and construction.

Jacob Banta's chief challenges with the *Niagara* involved designing a hull that could simultaneously support the heavy machinery amidships, offer the longitudinal stiffness needed to prevent hogging at the bow and stern, and achieve a high rate of speed—all without drawing more than seven or eight feet of water. Despite the *Niagara*'s burned and broken condition and mussel encrustation, divers today can still appreciate Banta's shipbuilding artistry and the quality of materials used. The size of the timbers and the large number of heavy iron fastenings, many of them expensive threaded bolts, are striking and contrast markedly with Great Lakes schooners of a similar vintage. The presence of such quantities of machined iron suggests Buffalo's growing industrial base and may help to explain why Charles Reed chose a Buffalo shipyard over his own facility at Erie.

The wreck reveals a durable ship, successfully designed to survive the almost routine array of accidents that plagued the Great Lakes, including collisions with sandbars, beaches, piers, ice, rocks, and other vessels, as well as to resist the internal stresses generated by the heavy steam engine and boilers and the occasional heavy seas. The ship's skeleton consists of longitudinal oak timbers, called futtocks or frames, 6 inches wide and about 12 inches deep at their thickest point. Frame thickness tapers from 12 inches at the centerline to 10 inches at the turn of the bilge and 6 inches at deck level. Nine inches of space separate each frame set. The consistent framing and well-shaped timbers testify to the high-quality workmanship evident throughout the vessel.

The *Niagara* features a split-frame, or open-joint, framing rare on the Great Lakes. Each frame set consists of two timbers separated by 3-inch-thick by 20-inch-long oak spacing blocks. The rationale behind this design is unknown. The spaces may have helped prevent dry rot by allowing better air circulation. With the addition of these spacing blocks, each frame set takes on an H-beam appearance, an arrangement that offered the dimensional strength of a larger solid frame set without the added weight or expense and was believed to result in a "stiffer" (a stronger and more rigid) hull.

A large, traditionally built wooden sailing vessel derives much of its longitudinal strength from a heavy backbone composed of a centerline keelson structure and a thick keel. Such an arrangement would provide poor support for the heavy low-pressure steam machinery; in addition, the substantial keel would cause a steamer to draw significantly more water. The *Niagara*'s centerline keelson is 14 inches wide by 10 inches deep and rests in a 1-inch-deep notch on top of the floor timbers or frames. Iron bolts pass through the keelson, frames 10 inches wide and 5 inches deep, and the ship's keel. The keel and keelson sandwich the frames, providing an evenly spaced series of integrated wooden structures 27 inches thick at the bottom of the hull.

Much of the longitudinal strength for *Niagara*'s machinery came from a pair of large arched keelsons located amidships that supported the engine and boilers. Fashioned from several pieces of heavy oak and fastened together with iron drift pins and huge threaded bolts, the two keelsons, spaced about 28 inches outboard from the vessel's centerline, run down each side of the steamer's bilge (the interior side of the ship's bottom). At their strongest point, directly beneath the engine, the keelsons measure 4 feet deep and 2½ feet wide. They are also notched into the frames, where they are attached by threaded iron bolts 1¼ inches in diameter and capped with 2½-inch washers and 2-inch square nuts. These keelsons diminish in size as they move away from the machinery at the center of the vessel, tapering down to single timbers about 10 inches deep.

Outboard from the heavy arched keelsons, two bilge keelsons composed of 14-inch wide by 10-inch deep oak timbers are bolted to the notched frames. Forward of the engine and boilers the *Niagara*'s five keelsons continue, with a strengthened centerline keelson 17 inches deep compensating for the reduced bilge keelsons. Although the arrangements varied considerably and grew more sophisticated with time, the use of multiple keelsons and a vestigial external keel became essential characteristics on all of the largest wooden vessels built on the Great Lakes.

Banta further strengthened the hull by adding at least eight or nine longitudinal thick strakes of unusually large dimensions to the interior of the frames at the turn of the bilge. Where the *Empire*'s thick strakes measured 8 inches by 8 inches, the *Niagara*'s were 10 inches by 10 inches, a more than 50 percent increase in internal volume. Using a control weight of forty-five pounds per cubic foot for seasoned American white oak, a 1-foot section of a thick strake from the *Empire* weighed about twenty pounds, whereas on the *Niagara*, the weight was just over thirty-one pounds. Over time, the weight of the hull would have increased substantially as moisture permeated aging timbers. The installation of the larger thick strakes added more than 20 tons to the weight of the *Niagara*'s hull. The builder attached the thick strakes to the ship's frames using between two thousand and three thousand threaded iron bolts, each ¾ inch in diameter and capped by 1⅝-inch square nuts, adding an additional 2 or 3 tons in weight. The unusually heavy thick strakes and the use of hundreds of heavy threaded bolts are features that archaeologists had not previously encountered on early Wisconsin shipwrecks. In some respects, the hull with its heavy bilge strakes and bolts exhibits characteristics associated with early steam warships and Atlantic deep-water steamers.[8]

Banta may have compensated for the weight of the thick strakes by using lighter planking inside and outside of the steamer. On the interior of the hull above the thick strakes, the builders attached 3- to 3½-inch thick ceiling planking to the frames. These milled oak planks are a treat to the modern eye. Up to 18 inches wide and in some cases exceeding 40 feet in length, these tightly fitted planks formed the ship's inner skin and added substantially to longitudinal hull strength. Based on the remaining wreckage, we can offer a rough estimate of about 125,000 board feet of oak for the ceiling planking, adding about 20 tons of weight. Sample measurements taken on the upper works of the hull revealed an exterior planking thickness of 2 inches. The exterior planking on the lower part of the hull may have been thicker, perhaps the same dimension as the ceiling planking. By contrast, George Washington Jones reportedly installed a 6-inch-thick ceiling and 6-inch-thick exterior planking on the *Empire*.

As with the *Empire*, the builders added additional strength to *Niagara* with arches. However, in the latter case, the arches are much lighter and integrated into the interior part of the hull in a manner also found on Great Lakes schooners. The *Niagara*'s arches overlay the ship's ceiling planking and thick strakes. The iron fastenings attaching the arches to the hull also pass through the ceiling planking and thick strakes, creating a stronger structure. Replaced or added in 1851, the arches are essentially an additional layer of ceiling planking and consist of 3-inch-thick oak boards of varying widths. Commencing below the turn of the bilge, the arches peaked at or near the points where the heavy paddlewheel shafts crossed the sides of the hull. Largely burned away, the arches had an estimated length of 118 feet, or about half of the *Niagara*'s length, and were perhaps 10 to 12 feet in height at the top of the arches, with a maximum thickness of slightly over 5 feet. Tightly fitted and fastened to heavier timbers, the arches added significant support amidships with only a modest addition in weight.

Rough contemporary, and possibly generic, drawings depicting the *Niagara* show no evidence of trusses or other strengthening features above the deck. The framing of the deck itself would have added strength, but as it did not survive the fire and the wrecking, that part of the architecture remains a mystery. The other major hull feature to survive is the detached stern assembly, which offers a partial measure of the ship's depth. The top of the sternpost is partially destroyed; however, the attached deadwood has a maximum length of 14 feet 5 inches. Both stern timbers are 16 inches deep and 6 inches wide, giving the overall stern timber assembly a thickness of 32 inches. The ship's missing rudder hung in heavy brass fittings called *gudgeon straps* that are strongly attached to the sternpost.

Jacob Banta's hull suggests something about how he and other shipbuilders of his time conceptualized the Great Lakes maritime environment. The exceptionally heavy construction provides physical evidence that supports David Stevenson's 1837 observations about the staunch and seaworthy appearance of Great Lakes steamers. The heavy hull weight came at a price. With a normal load, the *Niagara* would draw between seven and eight feet of water. When heavily loaded, it would draw even more. A rough estimate based on its gross dimensions suggests that the *Niagara*'s hull possessed approximately 3,400 tons of buoyancy. This represents the maximum weight, including its machinery and hull structure, that the hull could support without sinking. Every additional pound of wood and iron used in building the hull pressed it deeper into the water and reduced the margin available for cabins, machinery, fuel, and, critically, cargo. Despite the need for shallow draft, Banta designed a very heavy hull, one with much of the weight concentrated at or below the turn of the bilge—a feature that likely provided enhanced durability at critical points and that probably improved the stability of the high-sided, flat-bottomed steamer. With its vast size and the liberal incorporation of iron fastenings, the hull cost $30,000. This was money that proved well spent over the course of the *Niagara*'s career.

The most expensive component on any early steamboat was the engine, which was specially designed and manufactured for each boat. Engine building required a well-equipped foundry and skilled artisans and was a costly, labor-intensive process.[9] To power the *Niagara*, Reed turned to New York engine builder James P. Allaire. An associate of Robert Fulton, Allaire had purchased the Fulton Works after the famous man's death. In 1845, Allaire operated the oldest and finest engine works in the country. A 1980 PhD dissertation on the life of James Allaire states that none of the Allaire engines survived. Fortunately, with the documentation of the *Niagara*, this is no longer the case.[10] The ship was powered by an Allaire low-pressure walking beam steam engine. The "walking beam" refers to a large, iron diamond-shaped beam that towered above the ship and was balanced upon a giant A-frame. Walking beam engines appeared awkward and fragile but were in fact durable, safe, reliable, and, after a fashion, efficient at producing usable horsepower. They remained popular for coastal navigation into the early twentieth century. To foreign eyes, however, they seemed perversely American.[11]

Archaeological site plan of the *Niagara*

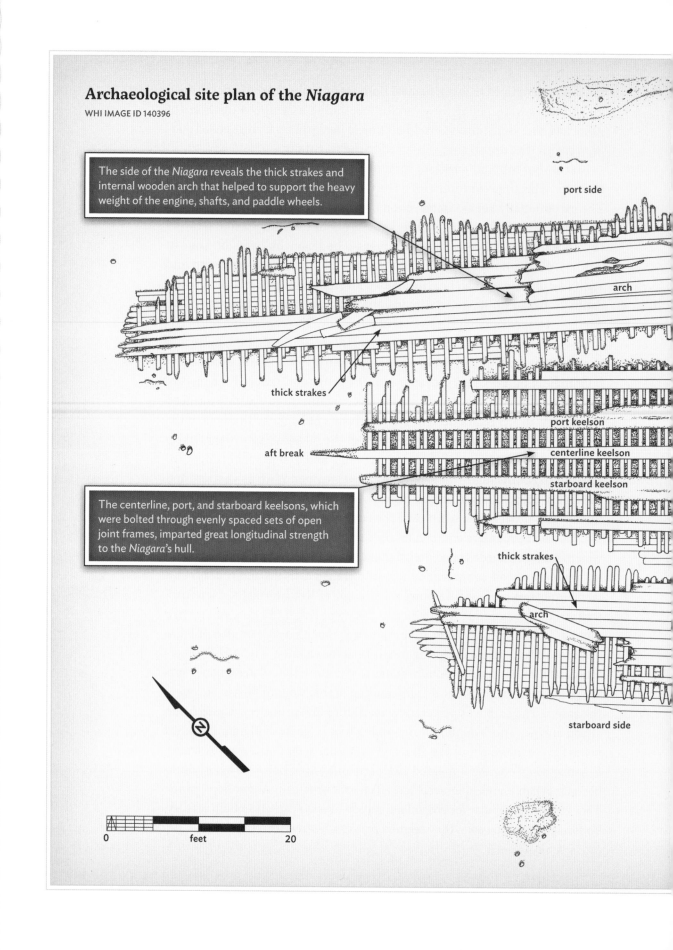

The side of the *Niagara* reveals the thick strakes and internal wooden arch that helped to support the heavy weight of the engine, shafts, and paddle wheels.

port side

arch

thick strakes

aft break

port keelson

centerline keelson

starboard keelson

The centerline, port, and starboard keelsons, which were bolted through evenly spaced sets of open joint frames, imparted great longitudinal strength to the *Niagara*'s hull.

thick strakes

arch

starboard side

0   feet   20

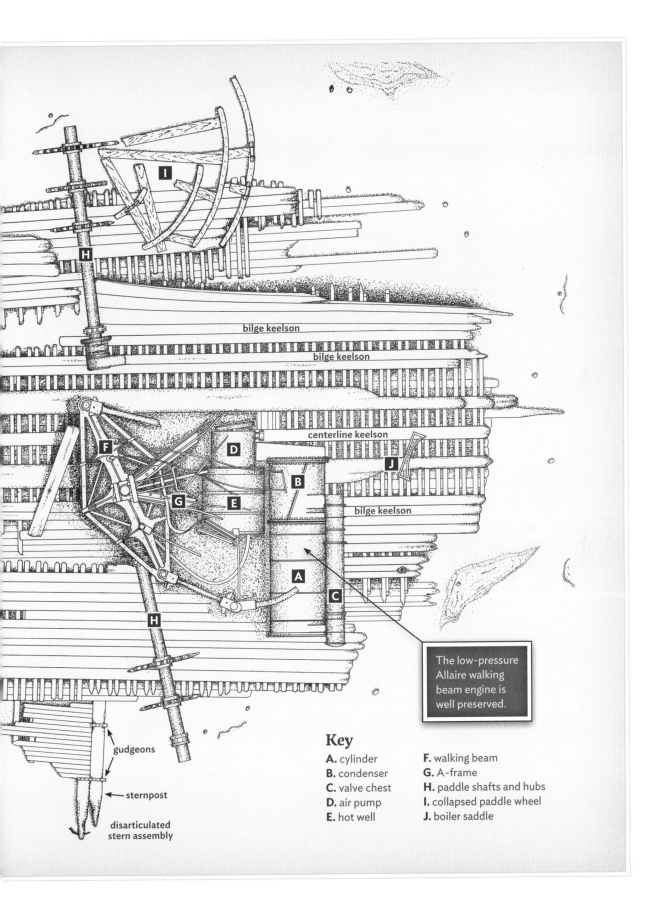

bilge keelson

bilge keelson

centerline keelson

bilge keelson

H

I

F

D

G

E

B

J

A

C

H

gudgeons

sternpost

disarticulated
stern assembly

The low-pressure
Allaire walking
beam engine is
well preserved.

Key

A. cylinder
B. condenser
C. valve chest
D. air pump
E. hot well

F. walking beam
G. A-frame
H. paddle shafts and hubs
I. collapsed paddle wheel
J. boiler saddle

Marine steam engines were individually crafted at large engine plants similar to New York City's Novelty Works. *HARPER'S NEW MONTHLY MAGAZINE,* MAY 1851

Fifteen years after the launch of the *Niagara*, the famed English naval architect Norman Russell described the genius and continuing ubiquity of the American walking beam engine at a session of the Institution of Naval Architects. Although it appeared at odds with what most English engineers considered desirable in a steamboat engine, Russell described the walking beam engine as "cheaper in construction, lighter in weight, more economical in management, less costly in repair, more durable, and better suited for high speed, than any of our own engines would be."[12]

The main part of the engine consisted of a vertical cylinder housing 14 feet tall and 6 feet in diameter. Enclosed in the housing was a cast iron cylinder 5 feet in diameter that traveled 10 feet up and down during every stroke of the engine. A large shaft transmitted the power expressed in the cylinders to the forward end of the walking beam, which rocked on trunnions supported by a heavy wooden A-frame.

The walking beam was 21 feet 4 inches long and 10 feet 6 inches high, typical of diamond-shaped beams used on vessels of the period. Composed of a cast iron interior frame and bounded by a forged iron band, the beam possessed considerable strength and resiliency. A beam of this size was undeniably heavy, perhaps 4 tons or more. While the weight was impressive on its own, its suspension on top of an A-frame 20 or more feet above the water's surface and high above the steamer's center of gravity multiplied the significance. The heavy beam would cause the ship to roll farther to each side when running sideways to the seas. It also, however, would substantially slow the speed of the rolling, making the ride for the passengers less jarring. Applying tables developed in the second half of the nineteenth century, the total weight of the engine, not including the massive boilers, was around 160 tons and provided ample counterbalance for the weight of the beam.[13]

On the opposite end of the beam, more iron shafts transmitted the engine's force to heavy iron cranks. Similar to the hardware found on a carousel horse, the cranks converted the up-and-down movements produced by the piston into circular motions that turned the paddlewheel shafts.[14] The cranks transferred the full power of the engine and, exposed to a variety of conflicting stresses from the paddlewheel and the engine, sometimes failed. This failure was a relatively common occurrence on the large steamers.

Each iron paddlewheel shaft measured 18 inches in diameter and nearly 30 feet in length and weighed more than 6 tons. Each paddlewheel consisted of three 6-foot diameter and 5-inch-wide hubs incised with twenty-five notches that held the 13½-foot-long spokes. Supported by curved iron strapping, the paddlewheels terminated in buckets consisting of wood planks 5 inches wide, 1¼ inches thick, and 9 feet 11 inches long. The overall diameter of the wheels was roughly 30 feet.

The walking beam was a low-pressure steam engine. While nominally safer because of the reduced risk of boiler explosions, the low-pressure engines were significantly larger and heavier than high-pressure engines. For example, although the engines on the *Niagara* and the *Empire* both had 10-foot piston strokes, the cylinder of the *Niagara* engine was 60 inches in diameter, and the *Empire*'s was just 35 inches. In general, however, the public in the 1830s and 1840s tended to prefer ships with low-pressure engines. During times of heightened competition for passengers, the owners of low-pressure vessels took pains to point out the superior sense of safety of embarking on their steamers.

High pressure or low, Great Lake palace steamers consumed staggering quantities of wood as fuel. In 1848, the *Niles National Register* reported that the *Empire* typically consumed six hundred cords of wood, and at times as much as seven hundred, traveling round-trip from Chicago to Buffalo. The report estimated that the ship consumed 234 acres of virgin forest and kept forty woodcutters employed.[15] Other knowledgeable observers from the time provide comparable figures. Based on this information, the *Niagara* probably burned about three cords of wood per hour, sufficient fuel to heat a modest-sized New England or midwestern home for an entire winter.

Over a season, its fuel consumption added up to more than 8 million board feet of timber, enough to build more than four hundred of those homes; and these are conservative estimates.

Supplying power to the *Niagara*'s engine was a boiler assembly 26 feet long, 27 feet wide, and more than 17 feet high. Despite mussel colonization, the heavy iron boiler survives in an excellent state of preservation. It consists of a single large firebox, 26 feet 8 inches wide, 7 feet 3 inches deep, and 11 feet high. Atop the firebox, two steam chambers extend up an additional 6 feet 6 inches. Attached to the firebox are three return fire tube boilers, each measuring 18 feet 10 inches long by 8 feet in diameter. The weight of the boilers, constructed principally from sheets of wrought iron plating, is unknown but substantial, perhaps 50 tons or more.

The weight of the machinery and its concentration in the center of the vessel explain the great effort of palace steamer builders in strengthening their shallow hulls. That the standardization of designs came slowly, if at all, and that individual builders continued to alter their designs during this period suggest that the problem proved difficult and success elusive. In the *Niagara*, and it appears in the *Empire*, the designers compensated for the weight of the machinery and weakness inherent in their vessels' lengths principally by adding mass to strategic parts of the hull. In the *Empire*, however, George Washington Jones went further with his external arches, probably the first used on the Great Lakes. In subsequent years, all of the largest steamers, including those designed by Jacob Banta, featured external trussing.

Steamboats were expensive to build, and the *Niagara* was no exception. The *Buffalo Commercial* provided readers with the following cost breakdown for the crack steamboat: $30,000 for the hull, $32,000 for the engine, $5,000 for fixtures, $8,000 for joiners, $3,500 for painting, $3,000 for upholstery, $4,000 for furniture, and $9,500 for other items, including anchors, chains, cables, boats, sails, blocks, cutlery, lamps, cooking utensils, and table linen. In total, the *Niagara* cost $95,000.[16]

The $23,000 spent on fixtures, joiners, painting, upholstery, and furniture reflected the final pillar of performance, the capacity of a palace steamer to project an image of progress of civilization over the frontier wilderness. The size, luxury, and speed of palace steamers not only captured the attention and affection of westerners but also garnered the nearly unanimous praise of cosmopolitan English travelers, otherwise so often contemptuous of American manners and customs. Several wrote with special regard for Great Lakes sidewheel craft. In the early 1840s, James Silk Buckingham punctuated descriptions of booming Chicago with references to Great Lakes steamboats, describing them as "some of the largest and finest steamboats that exist in the United States" and "of great solidity, equal to that of ships navigating the oceans."[17] Thomas Horton James, in 1845, explained to readers that when compared to the small size of British steamboats, the "magnificent" steamers of Buffalo were "very humiliating, and enough to make the Englishman blush for his inferiority."[18] In the early 1850s, Alexander Marjoribanks, examining the Lake Erie steamboat *Mayflower* built at Buffalo in 1848, declared it "magnificent beyond description."[19] Isabella Bird's fiery 1854 pen, however, better captured the *Mayflower*'s opulence: "My bewildered eyes surveyed a fairy scene, an eastern palace, a vision of the Arabian Nights. I could not have believed such magnificence existed in a ship. It impressed me much more than anything I have seen in the palaces of England."[20]

Chapter Four

STEAMING THE LAKES ON PALACE STEAMER *NIAGARA*

Nearly a year passed between the *Niagara*'s launching and its inaugural voyage. During the interval, Buffalo's finest artisans transformed a floating island of oak and iron into a frontier fantasyland. The *Niagara*'s first test cruise occurred on April 15, 1846. Despite choppy seas and a roughly operating steam engine, this shakedown cruise garnered the *Niagara* favorable notices for its speed (said to have exceeded fifteen miles per hour) and luxurious accommodations.[1] Less than a week later, under the command of Captain Thomas Richards, a pious and playful man who had emigrated from Wales and had worked for Charles Reed for several years, the *Niagara* embarked on what was described as a "pleasure trip" to Detroit. During the journey, Captain Richards, undoubtedly under orders from Reed, invited the entire Michigan legislature out for a pleasure trip on the new steamer. They responded with a unanimous official resolution to accept the invitation. The steamer concluded the journey by returning to Buffalo with a small cargo of flour and casks of ashes. It received complimentary reviews from the *Detroit Advertiser*, which noted of the *Niagara*, "We had been led to anticipate a most magnificent boat, but the reality far exceeded our highest anticipations."[2]

The *Niagara*'s next voyage illustrated the dilemma of the large palace steamer. Dispatched to Chicago, the trade for which it was designed, the heavy ship grounded at St. Clair Flats. Despite transferring cargo to a smaller steamer, the *Niagara* failed to make it through and was forced to return to Detroit, where it loaded a modest cargo of western bounty that

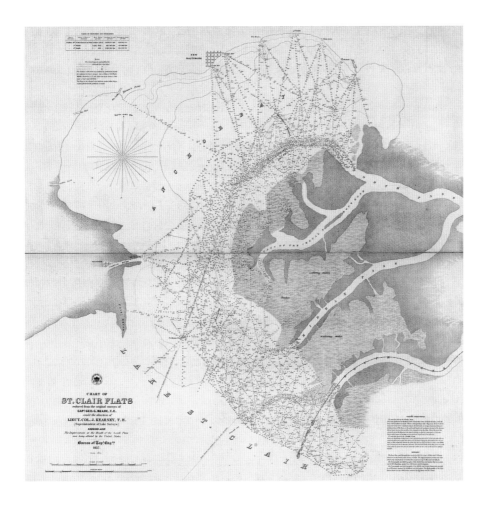

This nautical chart of St. Clair Flats from 1857 reveals the treacherous path between Lakes Erie and Huron. NOAA'S HISTORICAL MAP & CHART COLLECTION

included 25 tubs of butter, 16 barrels of deer skins, 483 hides, 725 barrels of flour, 731 pigs of lead, 3 boxes, and 4 chests of furs.[3]

Fortunately for Reed, the water level at the St. Clair Flats soon rose, following its annual cycle. On May 18, 1846, the ship surprised the people of Milwaukee, Wisconsin, arriving in great style with musical accompaniment and giant pennants flying. Although one almost never finds a negative review of a new ship, the *Niagara* made a genuine impression in Milwaukee, where it would become a favored ship. Crowds rushed to the pier for a look, prompting a reporter for the *Milwaukee Sentinel* to write:

> The *Niagara* is, in truth, a noble boat; well modeled, capacious, convenient, swift and most strongly built. Her after Saloon, on the upper deck, is superbly fitted and furnished and the accommodations for steerage passengers on the main deck are unsurpassed."[4]

The trip concluded successfully with the steamer returning to Buffalo carrying "a large load of freight and passengers" that included 61 tons of beef, 23 barrels of pork, 569 dry hides, and various allotments of furs, flour, wool, and sugar.

Luxury, size, and speed brought status to a steamer and to its owner, captain, crew, and passengers. The *Niagara* clearly possessed the first two characteristics and soon demonstrated the third. In early June, with Charles Reed on board, the *Niagara* set a record for a steamboat trip between Chicago and Buffalo, seventy hours (seventy-five hours counting port stops), returning to the home port thirty-five hours ahead of schedule.[5] The fast trips continued throughout the summer, with another record trip reported in late July.[6] Pageant and posturing were all part of the competitive world of pioneer steam on the Great Lakes. During this period, the *Niagara* engaged in a brief race with the *Empire*, which may have gotten the better of that encounter. The *Niagara* had to give way to avoid collision.[7] Captain Tommy Richards denied racing and claimed that his *Niagara* had gotten off course and merely ran alongside the *Empire* for a few moments.[8]

Autumn brought diminishing lake levels, with the *Niagara* grounding three times during an early October trip from Chicago.[9] The news was worse two weeks later: the *Buffalo Commercial* reported the *Niagara* running aground on the Skillagalee shoals in Lake Michigan and jettisoning four hundred barrels of cargo and part of its fuel to float free. The situation had been worse than the newspapers reported, and the ship's survival, much less with only minor damage, reflected its staunch construction and the extreme prudence of the officers. A deposition given by Captain Richards, filed before Michigan Justice of the Peace Charles M. O'Malley on October 23, 1846, recorded the ordeal in spare prose:

> After being out about two hours, it commenced to blow from the southward and Eastward fresh. The night being very dark and the weather squally with snow at about eleven o'clock, at which time we supposed that we were down nearly to Point Wagushanic [Waugoshance]. We could not see the point nor the light ship. We struck on the shoals of Skillagallee at which time there was due and necessary caution used and the boath [sic] of us were there on Deck and the Second Mate was heaving the lead at the same time.[10]

Blinded by bad weather in a dangerous part of the lake, Captain Richards could not determine *Niagara*'s position. Lost, the captain slowed the ship to

A C. Patrick Labadie drawing of the *Niagara* based on archaeological and historical sources.
C. PATRICK LABADIE

a crawl. Using a lead—a sounding device consisting of a regularly marked line to measure water depths, attached to a heavy weight loaded with sticky animal tallow that could pick up distinctive types of sand or pebbles—the *Niagara*'s officers slowly felt their way along the bottom. Although failing to find safe passage, their caution prevented a catastrophe when the ship eventually grounded.

On reaching port, Richards penned a quick note to his wife that offers a rare unguarded glimpse into the mind of a steamer captain. The opening of the letter embraced religion in underscoring the seriousness of the event: "I employ the first opportunity of writing you that you may know that I am yet preserved & praise Him who alone is able to save us when all human effort seems to fail." The body of the letter filled in many details not included in the official record. The grounding occurred within the context of a three-day storm of terrifying force. The violent seas created extraordinary stresses, which exposed weak joints in the new steamer's boilers. The resulting leakage of steam rendered the engine inoperable, leaving the ship nearly helpless. Like many other early steamers, the *Niagara* carried a single mast with fore and aft sails on the forward deck. Captain Richards attempted with little success to sail the huge craft through the stormy lake. The sails may have given the ship some added stability for a time, but they were utterly inadequate to propel such a heavy ship during a storm. The engineer succeeded in repairing the boilers and getting up steam, but violent seas made running the engine impossible. The storm eventually tore away the ship's tall smoke stacks and damaged the rudder, factors that contributed to the ship's grounding. Richards concluded the letter home with a poignant request:

> Louisa join me in sending thanks to Him who alone is able to bring us safely though the perils By Sea & Land if ever I felt my dependence on my maker it is at this time pray that He may continue to watch over & protect us from sickness & all the ills which mankind are subject to and above all that He may cleanse us thoroughly from all sin finally and together enjoy his presence forever. Give my love to all & believe me to be your affectionate Tommy (in haste).[11]

Richards remained the *Niagara*'s captain into its second season. In 1849, while commanding Reed's new palace steamer, the *Keystone State*, on its maiden voyage to Chicago, Richards fell ill and died, probably a victim of the cholera epidemic then raging through the Midwest.[12]

Late-season sailing is a factor in many of the wrecks and incidents described in this book. Lake weather deteriorates as summer gives way to fall and even more so with the coming of winter. The late season, however, offered many opportunities for shippers. Raw and processed material from the West needed to move east before the lakes froze over. Vessels remaining in service faced increased threats from the weather and, frequently, from lower lake levels. Reduced competition from other vessels and higher freight rates forced owners and crews to balance the increased risks with the prospect of high profits. The November 23, 1846, *Buffalo Commercial* recorded the *Niagara*'s arrival from Chicago and printed a list of its cargo that provides a good sense of these late-season opportunities as well as frontier exports:

> Chicago, 522 bbls. flour Kimberly, Pease & Co.—200 do. Heley – 101 green 17 dry hides N. Case & Co. 16 half bbls fish Gelston & Evens—14 sks ginseng 30 bbls flour H.M. Kinne 86 do. Weed & Co. – 31 do. G.B. Walbridge & Co.—1486 pigs lead D.N. Barney & Co. – from Detroit 544 bbls flour J.A. Armstrong.[13]

The trip concluded with the *Niagara* grounding at Buffalo's south pier, where it remained for several days.[14] The steamer made one last trip that first season, leaving Ogdensburg, New York, "with a heavy load of merchandise" on November 27.[15]

The *Niagara*'s first year of service proved successful and eventful. Making one round trip every two weeks, the *Niagara* carried thousands of passengers and a large variety of cargo, including furs, minerals, agricultural products, and consumer goods. The maiden season set the pattern for several years of cross-lake service. During the shipping seasons of 1847, 1848, 1849, and 1850, the *Niagara*'s schedule included fourteen to sixteen round trips between Buffalo and Chicago. The boat typically operated from the middle of April through the end of November, leaving on alternate Mondays from either Buffalo or Chicago. December through early April typically found the *Niagara* laid up in Buffalo, a time when it received substantial annual maintenance.[16]

New ships evoke an excitement from the public that usually fades quickly. Although newspaper references to the *Niagara* decline in number after the first years, the steamer retained a sense of newness and prestige during its first three or four seasons, as a fixture on the Buffalo–Chicago route.

Second-season highlights included a new captain in Levi Allen, a fresh paint job, more groundings, and a rumored steamboat race. At midcentury, the rapid development of towns and cities and active boosterism contributed to deep rivalries between regional communities, both large and small. In the case of Buffalo and Cleveland, the rivalry extended out onto the lakes. In the summer of 1847, with new steamers going into service and challenging current favorites, racing fever broke out. On July 18, the *Buffalo Morning Express* reported "another race brewing." Captain Gil Appleby of the new Cleveland steamer *Sultana* was offering a $5,000 wager against any steamboat on the lakes willing to race him from Buffalo to Chicago. The enthusiastic Buffalo paper gleefully reported that

> C. M. Reed, Esq. owner of several steamboats, is not inclined to remain quiet under the imputation and proposes to accept the wager, and put on his best boat, the *Niagara* against the *Sultana*. . . . We can afford to risk a hat on the race, which shall hang on the bowsprit of the *Niagara*, labeled "come and take me." If the editor of the (Cleveland) *Plain Dealer* is inclined to make a corresponding demonstration, being a chapeau on the bow of the fine *Sultana*, we have no objection. We pause for a reply.[17]

The Cleveland *Plain Dealer*'s editor could not resist tweaking the Buffalo press and tempered the *Plain Dealer*'s obvious enthusiasm for the race with scolding remarks about the foolishness of steamboat racing: "Nothing that we can now say will prevent this race, and therefore we shall say nothing, but we want it distinctly understood that we are opposed to racing, especially among steamboats with BOILER's in 'em." That being said, the *Plain Dealer* concluded, "Let the bets go! We hang our hat on the *Sultana*'s Stern, and in somebody else's language, we say 'Come and take it!'"[18]

Whether the race took place is unknown; however, the excitement generated helps to expose the cultural and economic forces shaping the palace steamer era. The captain and owner of the new *Sultana*, Gilman Appleby, was in direct competition with Charles Reed on the Buffalo-to-Chicago run. Although well appointed, the *Sultana* measured less than 800 tons and could not match the *Niagara* for size and probably not for speed. Racing or speculation about racing, however, offered free advertisement that conveyed the idea of luxurious and speedy steamers. While the public officially admonished steamboat racing, it also loved speed. The only other mention found for the *Niagara*'s 1847 season, however, is less glorious and reports its rescue from a reef at Bois Blanc Isle by the *Empire*.[19]

SHIPBOARD LIFE ON PALACE STEAMERS

Lake Michigan has preserved the essential remains of the *Niagara*'s nautical architecture and engineering. Although these remains allow us to understand better the relationships between maritime technology, the Great Lakes environment, and the economy during the era of pioneer steam, they speak less directly about human organization and behavior. A ship is a complex cultural object. Designed, operated, and inhabited by people, a ship possesses a spatial geography that reflects its intended mission, operational organization, and social hierarchy. Because of their excellent preservation, many Great Lakes vessels also preserve in place the material culture of shipboard life. This is most often true on schooners where the living spaces are located totally or at least principally below the main deck in areas that might remain intact when a ship sinks. Despite the destruction caused during the fire and sinking, the wreck of the *Niagara* once contained thousands of intact artifacts. Reportedly, the diver who rediscovered the wreck thoroughly stripped it. Guarding the wreck with a rifle, the diver and his companions removed intact cases of china, glassware, and any number of other items that could have added substantially to the collective understanding of life on a palace steamer. Today such activities are illegal in the Great Lakes but sometimes still occur.

The design of palace steamers made the complete preservation of cabins far less likely than on sailing vessels. The durability built into the hulls did not extend to their cabins. Constructed of lighter wood, principally pine, the cabins had to withstand abuse from wind, ice, and the occasional wave, but they were not designed for direct exposure to the water. When a large wooden passenger steamer sank, the buoyancy of pine decks and cabins frequently caused them to separate from the hull. Furthermore, as in the case of the *Niagara*, fires claimed many of these vessels, consuming the cabins, although not all the cabins' contents. On the palace steamers, when fires or other disasters compromised the arching and trusses that helped to hold the hulls together, these tended to break up. In sum, their design and construction materials combined with the violence of wreck events to make the survival of cabin spaces on large sidewheel passenger steamers unusual.

Personal accounts describing shipwrecks are common. No one who has survived a shipwreck, whether hardened salt or green traveler, forgets the experience. Furthermore, the legal and cultural context surrounding shipwrecks encourages the production of detailed accounts. The typical successful voyage, however, inspires few to write and fewer yet to publish. Fortunately, at least one *Niagara* passenger, Bishop Thomas Asbury Morris, whose observations opened the previous chapter, left a textured account of his 1848 voyage from Detroit to Chicago. His is the only detailed account by a passenger from the *Niagara*'s early years, and it offers an important and reliable look into life on a Great Lakes palace steamer and helps to place the ship into its larger social context.

Born in a frontier area of Virginia in 1794, Bishop Thomas Asbury Morris was a national figure in the Methodist Episcopal Church. His rather stout and bespectacled appearance belied a personal history of years of rugged frontier travel by foot, horse, canal boat, steamer, and railroad. Through those spectacles, Morris made keen observations and judgments (often moral) about his surroundings and his companions. As editor of the church organ *Western Christian Messenger* and a person with a sense of his own posterity, Morris kept a diary and frequently published accounts of his various travels.

At the beginning of July in 1848, Morris left his Cincinnati home on a journey to Milwaukee, where he would attend the meeting of the new Wisconsin Conference of the Methodist Episcopal Church. Reaching Sandusky by rail, he boarded a small steamer bound for Detroit. There he took passage on the *Niagara*, a choice preferable to the still rough overland stage and rail route across Michigan.[20]

Bishop Morris made several references to traveling by steamers in his published writings. Of these, however, he seems to have devoted the most detailed attention to his *Niagara* journey. As a physical object, the *Niagara* impressed Morris. He described in minute detail the elaborate life-sized carving that adorned the ship's bow stem where it projected a sense of frontier power and primitive nobility:

> Her forward keel was ornamented with a full-length figure of an Indian chief, with his flowing robe, raven locks, jewelry, and cap of feathers. He stood erect, with lofty bearing, his right foot in advance, a bow in his left hand, and a quiver projecting behind the left shoulder, his right hand firmly grasping the handle of a scalping-knife, with the point downward; and with his piercing eye fixed on some distant point, he seemed, by a most determined air, to say, "Follow me."[21]

Morris also lauded the ship's obvious strength and its twelve to fifteen mile an hour speed. During the tranquil early phases of the voyage, the passengers, the bishop reported, "scarcely heard her engines, or felt her move."[22]

Passing uneventfully through the St. Clair Flats, the *Niagara* steamed north on the clear and deceptively peaceful Lake Huron. Morris recounted a beautiful passage, one punctuated by his witnessing "one of nature's most beautiful exhibitions" in a cloud formation and an enormous multicolored rainbow as they passed by the notorious Thunder Bay. At Mackinaw, the hurried *Niagara* stopped for a half hour. Morris debarked to shop for "Indian curiosities," which he found picked over and of little interest. Of Indians, he saw many fewer than on a previous visit in 1844.[23]

The bishop had more to say about his fellow passengers and the organization of life on the great steamer. He described the *Niagara* as "a little world of itself." A person of keen eye, he noted the hierarchical separation of the passengers. On the lower deck, a large space with no staterooms and limited comforts, he observed crowds of people, mostly immigrants, "figuring in quaint costume," surrounded by their luggage and property.[24] Despite the growing nativism in America, Morris looked on the *Niagara*'s immigrant passengers with patronizing favor. He approved of their naïve enthusiasm for America and the careful ministrations of the wives and mothers in tending their flocks of children. These immigrants, at least, seemed to bode well for the nation's future.

At Castle Garden in New York City, agents for Charles Reed and the Erie Railway collected the immigrants off the ship and sold them railroad and Great Lakes steamer tickets to new homes across the Midwest. MOTT, *BETWEEN THE OCEAN AND LAKES*, 429.

Whether on crossing the Atlantic Ocean or the Great Lakes, immigrants offered high profits for ship owners, who packed them in as tightly as circumstances permitted. Charles Reed did well carrying immigrants to the West. By the 1840s, Reed's agents would meet new arrivals at the New York City docks, selling them westbound tickets along a railroad, canal, and steamboat route almost entirely owned outright or in part by Reed.[25] On a prosperous voyage, the *Niagara* might carry five hundred or more immigrants in addition to other cargo.

Bishop Morris, however, traveled in more comfort on the deck above, secure in its higher elevation and social status. The specific details of the *Niagara*'s upper cabin are unknown, but it followed a general plan common for the time and consisted of a series of staterooms surrounding a long salon where passengers ate, drank, and socialized. The wealthiest travelers and those traveling as families booked the staterooms, while others, usually men, occupied less private berths, made up by the stewards at the end of each evening's social activities.

The salon also provided a large tapestry for displaying luxury and cultural status and usually featured elaborate works of art, including murals, paintings, and intricate glassworks. The size of the *Niagara*'s salon is unknown; however, on the *Empire*, the salon was 220 feet long by 15 feet wide and divided by two folding bulkheads into three compartments: a ladies' cabin, a lounge with a piano and comfortable furniture, and a 130-foot long gentlemen's cabin. Dominating the center of the three compartments were long tables, a feature common to palace steamer cabins.[26] Although not as long, the *Niagara*'s salon doubtless featured décor of equal or greater quality. Unfortunately, neither Morris nor others left a clear description of the *Niagara*'s luxury adornments and their symbolism.

Fortunately, Morris exhibited a keener interest in people.[27] Close company with fellow cabin passengers inspired the bishop's normally turgid pen to reach near poetic heights, capturing in a few sentences the cosmopolitan and eclectic culture of the Midwest's urban frontier:

> In the cabin were men of leisure and pleasure with their families, seeking new sources of enjoyment—men of business, intent on its accomplishment—invalids traveling for health—peddlers of books and maps—tourists exploring new states—ministers and agents with ecclesiastical business, and smoking, loquacious politicians—some promenading the deck in solitude, some clustered together in social chitchat, others attracted by the sound of music and song.[28]

To pass the time, the passengers chose recreations that ranged from reading the Bible and praying in their cabins to carrying the "History of the Four Kings" (Morris's scolding euphemism for a deck of cards) up to the hurricane deck, a favored place to drink and gamble. Accounts of steamer travel frequently reference dancing. As Morris's first day on the *Niagara* passed into evening, the stewards cleared the tables from the ladies' cabin to make room for music and dancing. Although attending, Morris described the events that followed with a liberal mix of scorn, racism, and disapproval.

At an appointed time, probably around seven o'clock, a black musician picked up his fiddle and started with a lively tune that called many people to the open floor, where they "commenced running past each other, and facing about with a regular step." Morris derided the music as just a "screaking noise." To Morris's professed wonder, the unnamed musician cast a spell over the passengers and called dancers from all quarters, "boys and misses, dandies and flirts, men and women" to the floor.[29]

The dancing was part of the daily cabin routine carefully organized around eating and occasional stops to discharge and take on passengers and cargo as well as fuel. The day began with a leisurely breakfast served throughout the morning and continued with a noon lunch, a midafternoon dinner, and a heavy tea from seven to eight in the evening. Music and dancing followed the tea and continued late into the evening, augmented with a supper table set out at ten o'clock. This level of service did not continue throughout the palace steamer era—but it seems the standard during the early years.

The loquacious political kingmaker Thurlow Weed, traveling in the summer of 1847, described life on the steamers *Empire* and *St. Louis*, including the nightly dancing, with unrestrained enthusiasm. Food figured prominently in the experience. The *Empire*'s menu included large quantities of delicious Great Lakes whitefish and trout, freshly procured during the voyage. Weed's account of the stores consumed during the sixteen-day *St. Louis* excursions suggests few empty bellies. The meat and seafood list seems extraordinary: 16 quarters of beef, 9 calves, 22 lambs, 11 sheep, 18 pigs, 60 beef tongues, 3 barrels of corned beef, 2 barrels of corned pork, 40 hams, nearly 10,000 eggs, 600 chickens, 5 dozen turkeys, 450 pounds of bass and trout, 150 pounds of halibut, 125 live lobsters, 12 kegs of pickled lobsters, and 12 kegs of pickled oysters. In addition, Weed's party smoked 4,000 cigars and drank large quantities of alcohol. It was, after all, a time of shorter life expectancies.[30]

The Morris and Weed accounts reveal some of the deeper social and cultural significance of pioneer Great Lakes steamers such as the *Niagara* and the *Empire*. Serving as the principal way to cross the long distances to the expanding Midwest, the ships created new opportunities and facilitated the dreams of hundreds of thousands of men, women, and children of many races, nationalities, and social backgrounds. In the 1840s, when the typical lake component of the western journey began at one of the Lake Erie ports or Detroit, the duration of the steamboat voyage allowed a complex temporary community to form. This community reflected the evolving social composition and cultural tensions found in the rapidly growing Midwest.

In recent decades, historians have rediscovered the participation and significance of blacks in maritime America. Numerous studies have documented black seafarers as sailors, ship owners, maritime artisans, and fishermen. While the black presence on western riverboats is well recognized, their place on the Great Lakes is not. For the black musician playing the fiddle to Bishop Morris's chagrin, the *Niagara* provided food, shelter, employment, and a rare means to travel widely and with little fear.

Operating a palace steamer required herculean work on the part of the crew. By modern standards, the typical forty-person crew of a palace steamer is woefully small to operate a large ship and provide for the needs of five hundred or six hundred passengers. The passenger-to-crew ratio on luxury cruise ships at the beginning of the twenty-first century is frequently

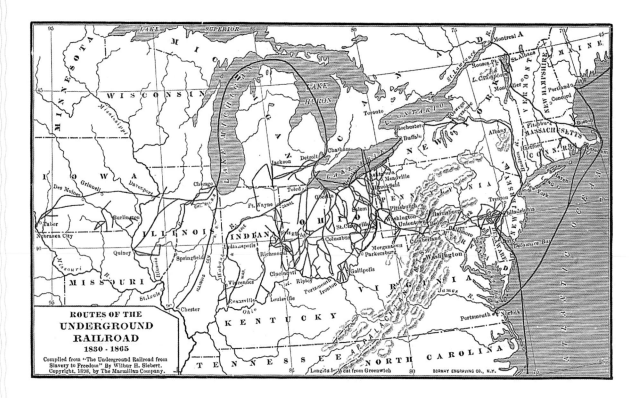

A map shows the route of the Underground Railroad through the Great Lakes, where former slaves escaping from the South found sympathy and aid on the *Niagara* and other steamers operated by Charles Reed. FOX, *HARPER'S ATLAS OF AMERICAN HISTORY*, 41.

as low as two passengers to one crew member, with large staffs dedicated to passenger service. On the *Niagara*, the figure was more often twelve or thirteen to one.

The domestic side of the palace steamer enterprise must have proven grueling, but it provided important opportunities for black workers who served as cooks, waiters, stewards, barbers, and musicians. A less enthusiastic description of steamer life, written in the early 1850s by Captain Lauchlan Bellingham Mackinnon of the British Royal Navy, described with humor his ejection from an armchair aboard an unnamed but crowded steamer by an enterprising black waiter. Bellingham had failed to tip the requisite twenty-five cents that the waiter collected as "rent" from the six to eight chairs under his control. Black workers also may have served in the engine department, stoking the greedy boilers with unending quantities of wood.[31] With a fueling stop requiring the rapid loading of sixty to one hundred or more cords of wood, the opportunities for hard labor on a palace steamer seem endless.

Black travelers, serving as domestic servants to wealthy passengers or sometimes traveling on their own, frequently joined the black workers aboard the steamers. Although the *Niagara* and its sisters sailed through free waters, that freedom did not extend to all on board. The young Isabella Lucy Bird, traveling on the palace steamer *Mayflower* in the early 1850s, found herself sharing a small ladies' cabin with a young southern woman, her two slaves, and five former slave women from Tennessee who had recently purchased their freedom.[32] The racial and social integration Bird described and the mixture of slave and free blacks on the ship raise important issues. Although few Great Lakes historians have incorporated race, an influential history of the Underground Railroad compiled in the 1880s describes with convincing detail the significant role of the Great Lakes ships, both sail and steam, in delivering slaves to safety in Canada. Among the most active in this enterprise was Charles Reed. His steamers, including the *Niagara*, offered floating havens where free blacks could get work and where slaves, passing as free workers, could cross into Canada.[33]

UNCERTAINTY ON THE WATER

One theme recurs throughout Morris's account: a concern for his own safety. The large number of recent and catastrophic steamboat accidents filled the bishop with fear. As evening fell on the first day of his journey on the *Niagara*, Morris noted the changing appearance of Lake Huron, which transformed "from bright sky blue to a somber purple." He continued, "As night fell upon us, grave thoughts intruded themselves—three hundred souls aboard, with only a few inches of timber between them and [a] sheet of water two hundred and twenty miles long and one hundred and seventy-five broad with the ordinary risk of collision, explosion, and fire."[34]

While Morris did not approve of drinking, gambling, and the dancing on general principals, safety rather than moral propriety seemed his greatest concern when the new captain of the *Niagara*, William T. Pease, fulfilled his hosting duties by energetically outdancing all the passengers. Morris offered this scathing rebuke: "Suppose, while he was capering about and measuring his steps by the motion [of the fiddler's] elbow, surrounded with the stamping noise of a crazy multitude, the boat had suddenly taken fire, ten miles from land, who, amidst the darkness and peril of the hour, would have been responsible for the souls aboard, and prepared to meet the emergency with discretion?" The straight-laced Bishop considered Captain Pease's behavior an act of "rudeness . . . toward the civil part of the passengers, who had gone aboard with reasonable expectation of safe conveyance."[35]

Midway through the journey, a peaceful Lake Huron gave way to tempest on Lake Michigan. An easterly storm carrying rain and high winds prevented the steamer from landing at Manitou Island, and for twelve hours the boat "was rolling and pitching," causing widespread seasickness. The conviviality of the ladies' cabin while leaving Detroit gave way to a nearly empty breakfast table. Near Sheboygan, Wisconsin, the wind died, and as the *Niagara* approached the town on a glassy lake, Morris was greeted with the "appalling spectacle" of the charred keel of the propeller *Phoenix* that had burned the year previously, claiming nearly three hundred victims. One more reminder of the dangers of lake travel, the exposed wreckage of the steamer *Boston*, destroyed when driven ashore during a nighttime gale, greeted Morris on reaching Milwaukee. He sagely concluded, "These manifest tokens of destruction strongly admonish the passerby of the perils of the lake."[36]

The 1848 season continued the pattern of frequent groundings for the large steamers, including the *Niagara*. Every incident cost time and money and sometimes led to serious disaster. These groundings had political as well as economic ramifications. An editorial in the *Milwaukee Sentinel* railed against President James Polk's dealings with the Great Lakes:

> The St. Clair Flats—Mr. Polk's "Farm" on the St. Clair Flats seriously obstructs navigation again this spring. The *Niagara* both in coming up and going down, lay aground there some 24 hours. The *St. Louis* and *Hudson* were also detained at the same spot. . . . Few steamers, or vessels indeed, from this Lake, carrying loads, escape detention on Mr. Polk's "farm." This tax thus levied upon our Lake Commerce is a heavy one— Every hour lost, and every dollar paid at the flats, is a charge upon the producers and consumers of the West. Mr. Polk's constitutional scruples cost Wisconsin many thousand dollars annually. How long will Wisconsin submit to the imposition?[37]

Polk, a slave-owning Tennessee Democrat, had vetoed major legislation to fund improvements to harbors and channels across the Great Lakes. The political discord surrounding Great Lakes navigation resulted in many northwestern businessmen leaving the Democratic Party for the Whigs during the late 1840s. The ascendance of the Buffalo lawyer Milliard Fillmore as the last Whig president in 1850 paved the way for a huge $2.2 million appropriation for rivers and harbor improvements in 1852.[38]

The transportation revolution that swept the Midwest and the nation throughout the antebellum period increased the speed and reduced the cost of travel. By 1848, the cost of a trip from Chicago to Buffalo on the *Niagara*, according to one source, had fallen due to increased competition to only ten dollars.[39] Steamboat travel, however cheap and comfortable, had its hazards, as the public was regularly reminded. The only other reference to the *Niagara*'s 1848 season recounts the drowning of an unidentified male passenger who fell from the steamer's gangplank while going ashore in Cleveland.[40]

Evolving Travel Landscape

Except for coverage of its loss in 1856, detailed information about the *Niagara* became sparse after 1848. The highlights of the vessel's career detailed below, however, hint at the broad changes sweeping the midwestern transportation landscape and offer glimpses of the day-to-day challenges, dangers, and successes associated with operating a palace steamer.

In 1849, a deadly cholera epidemic, the one that likely killed the *Niagara*'s first captain, Tommy Richards, ignited fear and depressed commerce throughout the nation. Yet, the year proved profitable for Charles Reed's *Niagara*.[41] With Captain Pease in command, the vessel continued on the Buffalo-to-Chicago run. An advertisement in the April 14 *Buffalo Courier* gave notice of the steamer's early season trips and traded on the ship's well-known durability by advising travelers: "As ice is apprehended in the Straits, passengers would do well to take this strong and reliable boat."[42] On August 11, the *Niagara* delivered 650 barrels of bulk freight, 40 tons of merchandise, and 550 passengers (mostly Norwegian and Dutch emigrants) to Milwaukee.[43] Such trips made the *Niagara* and its captain favorites in Milwaukee.[44] On September 3, the *Niagara* arrived at Buffalo with $300,000 worth of furs and buffalo robes, then the most valuable cargo ever landed at that port. The newspaper covering the story credited the remarkable achievement to the new Illinois and Michigan Canal, which successfully diverted traditional Mississippi River trade to the Great Lakes.[45] The year closed with foul November weather that damaged several steamboats, including the *Niagara*, which limped into port with several paddle buckets torn away.[46]

Niagara's 1850 season foreshadowed the changing organization of Great Lakes transportation. While the ship continued its schedule of alternate Monday departures from Buffalo and Chicago, an expanding national railroad network was moving west. Bit by bit, passenger trains were taking over travel to the major Great Lakes ports. In the beginning, the new railroads fed the steamers by bringing more travelers to the eastern lakes, where they would complete their westward journey by ship. In 1850, the *Niagara* called at Milwaukee thirteen times and landed nearly four thousand passengers.[47] After the 1850 season, the ship underwent a major refitting that included

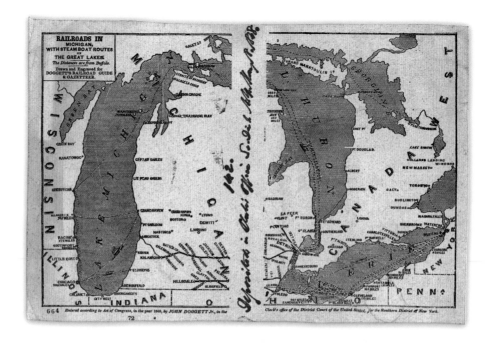

John Doggett's 1848 map shows the principal steamboat routes on the Great Lakes during the early years of the *Niagara*'s career. LIBRARY OF CONGRESS GEOGRAPHY AND MAP DIVISION, G4111.P3 1848 .D6

shipwrights installing new arches to strengthen the ship's hull. Newspapers speculated, correctly, that the steamer's future might lie in collaborating with the railroads.[48]

For Charles Reed, the railroads at first offered more opportunities than challenges. He invested early in rail and apparently saw the inevitable triumph of rail over steamboats in the passenger trade. In 1851, when the New York and Erie Railroad opened, the refurbished *Niagara* joined a flotilla of Buffalo steamboats that traveled to Dunkirk, New York, for a celebration. The gala featured dignitaries that included President Fillmore and US Senator Daniel Webster.[49] That year, Reed chartered the *Niagara*, *Empire*, and *Keystone State* to the railroad for a reported $65,000. Each carrying about sixty passengers per trip, a fraction of their capacity, the three steamers (now second-class boats) provided connecting service between the railroad's terminals at Dunkirk and Detroit.[50]

In the summer of 1851, the *Niagara*'s former first mate Fred S. Miller became the steamer's final captain. Late in the 1851 season, the *Niagara* was involved in a serious accident. While steaming down the Detroit River, it collided with the anchored brig *Lucy A. Blossom*. The brig, carrying ten thousand bushels of corn, sank quickly but not before crashing into another sailing vessel and tearing away part of its headgear. The staunch *Niagara* sustained no damage and continued on the Detroit–Dunkirk route through the 1853 season.[51]

NEW YORK AND ERIE RAIL-ROAD---TIME TABLE.

EXCURSION TRAIN.

May 14th, 15th, 16th and 17th, 1851.

May 17. Train.	STATIONS.	Dist. from NYork	May 14. 1st Train.	May 14. 2d Train.	May 17. Train.	STATIONS.	Dist. from NYork	May 14. 1st Train	May 14. 2d Train.	May 16. 2d Train.	May 16. 1st Train.	STATIONS.	Dist. from NYork	May 15. 1st Train.	May 15. 2d Train.
7.00 PM	Ar. New-York. Dep	--	6.00 AM	6.00 AM	1.42 PM	Dep. Shohola, Dep.	117	11.51 AM	12.49 PM	6.34 PM	5.57 PM	Ar. Elmira, Dep.	283	6.30 AM	6.35 AM
5.23 PM	Ar. Pier, Dep	24	8.00 AM	8.05 AM	1.32	· Lackawaxen, ·	121	12.01 PM	12.50	6.28	5.52	· Junction, ·	287	6.36	6.42
5.20	Dep. Piermont, ·	25	8.04	8.09	1.18	· Mast Hope, ·	126	12.14	1.05	6.18	5.42	· Big Flats, ·	293	6.45	6.52
5.12	· Blauveltville, ·	29	8.11	8.17	1.05	Narrowsburg, ·	132	12.24	1.25	6.05	5.30	· Corning, ·	301	7.02	7.17
5.03	· Clarkstown, ·	33	8.21	8.30	12.35	· Nobody's T't, ·	137	1.03	1.36	6.53	5.22	· Painted Post, ·	302	7.04	7.21
4.58	· Spring Valley, ·	35	8.27	8.37	12.26	· Cochecton, ·	141	1.12	1.45	5.38	5.08	· Addison, ·	312	7.17	7.36
4.56	· Monsey, ·	37	8.25	8.45	12.15 PM	· Callicoon, ·	146	1.23	1.57	5.30	5.01	· Rathboneville, ·	317	7.25	7.45
4.50	· 15 mile Turnout	39	8.40	8.52	12.00 M	· Fremont, ·	153	1.59	2.14	5.18	4.49	· Cameron, ·	324	7.55	7.58
4.43	· Suffern, ·	42	8.47	9.00	11.36 AM	· Equinunk, ·	163	2.02	2.40	4.58	4.27	· Canisteo, ·	337	7.55	8.15
4.39	· Ramapo, ·	44	8.50	9.04	11.23	· Stockport, ·	169	2.15	2.54	4.52	14.22	· Hornellsville, ·	342	8.10	8.35
4.30	· Sloatsburg, ·	45	8.52	9.08	11.13	· Hancock, ·	174	2.25	3.05	4.35	3.56	· Almond, ·	346	8.26	8.50
4.31	· Shultz' Turnout	48	8.58	9.15	10.58	· Hale's Eddy, ·	182	2.41	3.21	4.24	3.46	· Baker's Bridge, ·	350	8.37	9.02
4.23	· Monroe Works, ·	52	9.06	9.24	10.48	· Deposit, ·	187	2.56	3.37	4.05	3.25	· Andover, ·	358	9.02	9.32
4.18	· Wilkes, ·	54	9.12	9.31	10.24	· Summit, ·	194	3.16	3.57	3.45	3.00	· Genesee, ·	367	9.18	9.48
4.12	· Turner's, ·	57	9.19	9.39	10.03	· Susquehanna, ·	202	3.43	4.23	3.39	2.52	· Scio, ·	371	9.24	9.55
4.05	· Monroe, ·	59	9.24	9.45	9.42	· Great Bend, ·	210	4.00	4.42	3.31	2.42	· Phillipsville, ·	375	9.30	10.03
4.00	· Oxford, ·	62	9.31	9.52	9.32	· Kirkwood, ·	216	4.10	4.54	3.25	2.35	· Belvidere, ·	379	9.37	10.10
3.55	· East Junction, ·	63	9.34	9.56	9.25	· Windsor Road, ·	220	4.17	5.02	3.18	2.25	· Friendship, ·	383	9.44	10.18
3.54	· West Junction, ·	64	9.35	9.57	9.17	· Binghamton, ·	225	4.28	5.12	2.58	2.05	· Cuba, ·	390	10.02	10.32
3.51	· Chester, ·	65	9.37	10.06	8.57	· Union, ·	233	4.43	5.27	2.42	1.49	· Hinsdale, ·	397	10.20	10.47
3.38	· Goshen, ·	70	9.47	10.12	8.47	· Campville, ·	240	4.53	5.38	12.26	1.34	· Olean, ·	404	10.34	11.10
3.30	· New-Hampton, ·	74	9.56	10.22	8.37	· Owego, ·	247	5.40	5.55	2.08	1.11	· Allegany, ·	408	10.42	11.17
3.22	· Middletown, ·	77	10.11	10.31	8.22	· Tioga, ·	252	5.19	6.07	1.36	12.38	· Great Valley, ·	421	11.14	11.49
3.08	· Howell's, ·	81	10.20	10.41	8.13	· Smithboro', ·	256	5.26	6.17	1.12	12.12 PM	· Little Valley, ·	431	11.35	12.13 PM
3.00	· Otisville, ·	85	10.31	10.52	8.10	· Barton, ·	259	5.39	6.24	12.52	11.53 AM	· Albion, ·	438	11.55	12.33
2.45	· Shin Hollow, ·	92	10.45	11.08	7.58	· Waverly, ·	266	5.41	6.36	12.24 PM	11.25	· Dayton, ·	447	12.25 PM	1.00
2.30	· Delaware, ·	96	11.04	11.49	7.50	· Chemung, ·	270	5.49	6.48	11.53 AM	10.51	· Forestville, ·	461	12.57	1.32
2.03	· Rose Turnout, ·	108	11.27	12.16 PM	7.40	· Wellsburg, ·	276	5.58	7.00	11.30 AM	10.30 AM	Dep. Dunkirk, Ar.	465	1.15	1.50
1.52 PM	Dep. Middaugh's, ·	112	11.37	12.27	7.30 AM	Dep. Elmira, Ar.	288	6.09	7.11						

During the early 1850s, several new steamboats dwarfing the *Niagara* in size appeared on the Great Lakes. Steamers such as the *Western World* and the *Mississippi*, operated by the railroads, surpassed the *Niagara* by a hundred feet in length, nearly doubled it in registered tonnage, and exceeded its top speed by five miles per hour. The larger vessels, however, caused the holdups and accidents on the St. Clair Flats to increase, with costs associated with collision and grounding exceeding $500,000.[52]

The *Niagara* began the 1854 season contracted with the New York and Erie Railroad to run between Dunkirk and Toledo with stops at Erie, Cleveland, and Sandusky.[53] That year, Charles Reed initiated Reed's Chicago & Lake Superior Line, which offered service from Chicago to the shores of Lake Superior, and shifted his *Niagara* to the new service. On July 22, 1854, the *Cleveland Morning Leader* reported the deaths of several *Niagara* passengers from cholera during the previous Chicago-to-Mackinac trip.[54] Bad luck continued later in the season, when the *Niagara* required $2,000 in repairs to stem a major leak.[55]

The *Niagara* was a strong ship. Storms, ice, and collisions with other ships, piers, and rocks, as well as repeated groundings, proved that beyond doubt. After ten years of service, the steamer enjoyed an excellent reputation for strength, seaworthiness, and cabin service. Despite constant changes in the organization of passenger travel across the Great Lakes region and the

New York and Erie Railroad time table. MOTT, *BETWEEN THE OCEAN AND THE LAKES*, 97.

The era of the Great
Lakes palace steamer
climaxed with the
building of immense
vessels such as the
Western World, which
were roughly twice the
tonnage and substan-
tially faster than the
Empire and *Niagara.*
WHI IMAGE ID 63142

construction of palace steamers of nearly twice the *Niagara*'s tonnage, the
steamer remained profitable and popular. In large part, the *Niagara*'s continued success reflected the entrepreneurial skill of its owner Charles Reed,
whose grasp of the shifting transportation landscape and technologies had
few equals in the Midwest.

The completion of the Lake Huron segment of the Northern Railway on
January 1, 1855, signaled another important change in the transportation
geography by connecting Toronto with Collingwood, Ontario—a small outpost on Lake Huron's Georgian Bay. The railroad drastically reduced the
time required to travel from New York, New England, and eastern Canada
to the booming cities of Chicago and Milwaukee.[56] Travelers covered most
of the journey over land on the rails but either began or concluded their
trip with an overnight voyage on a lake steamer. To provide that service,
the Northern Railway turned to Reed, whose ships had dominated the
Buffalo-to-Chicago passenger trade for two decades. For the businessman
who deployed his steamers like chess pieces, this strategic move continued
to keep his fleet prosperous despite an overall decline in the number of miles
passengers traveled on the Great Lakes steamers each year.

The Northern Railway quickly proved a financial failure; high costs and
a depression led to its absorption by another railroad in 1858. Nevertheless,
during the first years, Reed's prudently managed steamers turned profits,
particularly through railroad subsidies. For example, the railroad reportedly
agreed to pay Reed $20,000 per ship for the *Niagara, Keystone State, Loui-
siana, Buckeye State,* and *Queen City* for their use during the 1856 season.[57]

Palace steamers retained a cachet that the railroads did not have. The well-maintained *Niagara*, despite its age, gained high marks from summer travelers who lauded it for having an "extensive and beautiful saloon, comfortable state-rooms and a table supplied with everything the appetite can desire." The comfort combined with the scenery made the Chicago-to-Collingwood trip, according to a Reed Line advertisement, the "most romantic, pleasant, and popular route ever established in the Union . . . the route for the Business Man, the Tourist, and the Pleasure Seeker." The ad also proclaimed it the "cheapest route to the East . . . and the only Route by which the tickets include staterooms and meals."[58] Although a financial failure for the railroad, the route attracted substantial numbers of passengers from Canada, New England, New York, and the Midwest. Long paid for, well maintained and managed by Reed, and competently commanded by the experienced Fred Miller, the *Niagara* might have continued to operate profitably for many years beyond the end of the palace steamer era had disaster not intervened.

Chapter Five

DEATH, REFLECTION, AND CHANGES IN DESIGN

As revealed in the writings of Bishop Morris, the possibility of disaster and death when traveling by steamer on the Great Lakes was difficult to ignore. After two decades of highly publicized catastrophe on ocean, river, and lake, by 1840, Americans could count nearly two thousand lives lost in various steamboat wrecks, collisions, explosions, and fires. Over the next twenty years, Great Lakes steamboats contributed substantially to increasing the death toll.

In so many respects, the sidewheel steamer *Niagara* was an admirable and safe vessel. Yet no amount of strong wood, iron fastenings, or clever design could address one particular danger, and that was fire. For centuries, the maritime world held a special dread for uncontrolled fire. During the age of sail, fires were rare, at least on civilian vessels, but when they occurred, the results were terrible. On the high seas, a large fire could easily mean certain death. The application of steam to navigation, however, transformed fire from a rarity to a common event—one that killed thousands of passengers and crew during the nineteenth century. With the huge fireboxes heating the steam boilers; sparks and cinders belching from the stacks; gaily painted pine cabins; long, open, and well-ventilated salons; and oil lighting, steamboats seemed designed to burn fast and hot.

Fires on the Great Lakes passenger steamers the *Erie* in 1841, the *Phoenix* in 1847, and the *G. P. Griffith* in 1850 alone claimed more than six hundred lives (170, 150–200, and 275, respectively). Scores of other lives were lost in many smaller fires and hundreds in other steamboat accidents.

Constant vigilance offered the most effective defense against fire on a palace steamer. Crews maintained a sharp lookout for small fires, which were commonly started by hot cinders discharged from the stacks, overheating stoves, or careless passengers. Buckets of sand strategically placed throughout the ship and used to smother flames were the first line of defense against an actual fire. Should the flames spread, the crew could deploy fire hoses and use fire axes to chop away burning structures.

On the western rivers, steamboat accidents killed a reported 1,155 persons between 1848 and 1852. In 1838, and again in 1852, the US Congress passed steamboat safety laws to curb the national steamboat carnage. The effectiveness of these laws, especially the second, continues to be a matter of scholarly debate. On the western rivers, steamboat boiler explosions dropped by more than half, and the number of lives lost fell by more than three-quarters during the first four years under the 1852 law. Steamboat accidents and deaths, however, climbed again during the second half of the 1850s.[1]

The 1852 federal law established the Steamboat Inspection Service and minimum standards for firefighting equipment.[2] Vessels the size of the *Niagara* were required to carry five fire buckets and one axe for every 100 tons burden. By these standards, the *Niagara* would have carried about fifty-five buckets and eleven axes. The law also required the carriage of three large, strategically located pumps, two of them manually powered, and long fire hoses that could be used to stop or slow the spread of flames.

The 1852 law required a vessel of *Niagara*'s tonnage to carry four boats, including one made of fireproof metal (if available) and "a good life-preserver, made of suitable material, or a float well adapted to the purpose, for each and every passenger." The *Niagara* carried five boats—two smaller boats high on the hurricane deck and a large boat and two good-sized quarter boats at the stern, none of them fireproof. The definition of what counted as a life preserver was vague. Chairs, stools, and doors—anything that floated and could easily be cast overboard—might qualify. The *Niagara* carried stacks of planks on the forward deck that fulfilled the law's life preserver requirements. It may not have carried personal life preservers. If that was the case, it was an unfortunate oversight, as their presence is credited with saving multiple lives when the 310-foot long palace steamer *Northern Indiana* burned on July 17, 1856. Another Bidwell & Banta–built steamer, the *Northern Indiana*, carried roughly 150 passengers and crew. The death toll in that disaster is unclear but was probably fewer than thirty.[3]

THE *NIAGARA*'S FINAL VOYAGE

If any of the two hundred passengers waiting to board the *Niagara* at Collingwood, Ontario, on the afternoon of September 22, 1856, had been well informed, they might have had some reasons for concern. The burning of the *Northern Indiana* was recent memory, and instead of their expected passage on the *Keystone State*, a 288-foot long, 1,400-ton palace steamer built for Charles Reed by Bidwell & Banta in 1849, they were being ushered aboard the older and substantially smaller steamer *Niagara*. Just days earlier, a September storm, described as one of the worst in memory, hammered the northern areas of the lakes, sinking several schooners and damaging many other ships. On the *Keystone State*, brutal seas staved in parts of the luxurious cabin and opened a leak that nearly sent the steamer to the bottom. To top it off, at the beginning of the *Niagara*'s previous voyage, the crew discovered a letter threatening to burn the *Niagara* "that very night." The captain and crew made a careful inspection of the steamer and found nothing amiss.

On the other hand, though older and smaller than the *Keystone State*, the *Niagara* was well known, time tested, and properly maintained. Its officers were skilled and experienced, with Captain Fred Miller having commanded the vessel for five years and served as its first mate before that. The clerk, George Haley, and the chief engineer, J. Leonard, had served on the *Niagara* from the maiden season in 1846. Although the regular chief engineer was not

on board for the unscheduled trip, his replacement also knew the steamer well. Given the vessel's history, the passengers boarding the *Niagara* had many reasons to feel confident that they would reach their destination in safety as well as in comfort.

Influential people boarded the *Niagara* that Monday afternoon, the most prominent among them John B. Macy, a former US Congressman from Fond du Lac, Wisconsin. A towering, corpulent man of great energy, Macy was returning from an aborted business trip to Lake Superior. Many families trooped aboard. Harvey Ainsworth was bringing his entire family—father, wife, sister-in-law, and three children—from Vermont to a new farm in Baraboo, Wisconsin. Martin Houghton, his wife, and their three children were traveling down from Westport, Canada. From St. Lawrence County, New York, Hugh Kennedy escorted his wife and daughter. L. Mattice, an engineer on the Fox River improvement project, was returning west with his wife and three children. In the steerage cabin, Mr. and Mrs. Chalmers shepherded eight children who ranged in age from infant to sixteen. Along with families came numerous men traveling on business, among them the cool-headed C. D. Westbrook of Green Bay. Several of the single travelers came from Canada, many from New England and New York, and a minority from the midwestern states. Carefully stowed in the hold and on deck were several wagons, twenty-one horses, and 105 tons of baggage and cargo.[4]

Marine artist Robert McGreevy shows the *Keystone State*, painted with Reed Steamship Line colors, steaming off the tip of Michigan's thumb prior to her layup in 1857. ROBERT McGREEVY

Under Captain Miller's careful watch, the *Niagara* left Collingwood at three o'clock in the afternoon, passed through the Georgian Bay, and stopped briefly at Mackinac to pick up six cabin passengers. At ten the following morning, the *Niagara* called at Two Rivers, Wisconsin, followed by quick calls at Manitowoc and Sheboygan, where the crew offloaded several wagons and steerage passengers. Most of the cabin passengers, however, continued on board, bound for the larger ports of Milwaukee and Chicago.[5]

According to Captain Miller's official statement filed in Port Washington, the *Niagara* left Sheboygan at two o'clock in the afternoon. Captain Miller then settled into his cabin for much-needed sleep. Two hours later, he awoke to find his ship on fire. Miller reported that he assumed the fire was located near the aft stove pipe—an observation that suggests this was not the first fire he faced during his years aboard the *Niagara*. Eyewitness reports, however, indicate that the fire began forward of the starboard paddlewheel. This would have placed it near the firebox of the boiler, the hottest part of the ship's interior. Captain Miller and his officers responded quickly, turning toward the Wisconsin shore a few miles to the west and deploying the fire pump and hoses.[6]

The fire's location and rapid spread made it impossible for the officers to reach the passengers clustering around the lifeboats. High off the water, untrained passengers attempted to launch two small boats from the hurricane deck. The first boat capsized. As a more well-ordered group lowered the second boat, filled with women and children, a panicking John Macy pushed his way to the hurricane deck. Screaming "We are lost!" Macy jumped from the *Niagara* and dropped an estimated seven feet into the heavily laden craft. The great weight of the corpulent politician tore the falls (lowering tackle) out of the lifeboat's stern, dropping the entire group into the lake and drowning them all.[7]

The news of Macy's death rippled through Wisconsin newspapers and took on a rather strange twist. The well-respected Macy was an ardent spiritualist. According to N. P. Tallmadge, a former governor of the Wisconsin Territory and Macy's neighbor, the congressman's ghost appeared separately to two of Tallmadge's daughters on the night of the sinking and before the news reached Fond du Lac. Both daughters reported Macy's death by drowning, and one was influenced to write the words "*Niagara*—Drown by the upsetting of the small boat."[8] In an apparent case of vanity beyond the grave, Macy's ghost failed to confess his role in causing so many deaths.[9]

The fire spread rapidly and within minutes disabled the engine and the pumps, leaving the vessel helpless. At this point, Captain Miller turned his attention to the safety of his passengers. Working his way aft, he began breaking off stateroom doors and tossing these and other buoyant articles over the side to provide floats. Miller remained on the deck until forced by the fire to take refuge inside the port paddlewheel, near the surface of the lake. With the wind blowing from the south and the vessel pointed to the west, the port side of the ship offered some respite from the flames. More than 160 years after the accident, the archaeological remains reveal the pattern of the fire, with the starboard side heavily charred and the port side largely untouched.

Passenger C. D. Westbrook left a corroborating account of the fire. In a cabin near the front of the ship when the fire broke out, Westbrook had quick access to the plank life preservers stored at the bow. As he prepared to abandon the steamer, he heard calls for help coming from the pilothouse. He responded by leaving the deck and assisted the wheelsman to steer the burning steamer toward the shore. Minutes later, after the engine stopped, Westbrook tossed a plank in the water and abandoned ship. Several passengers flailing in the water seized the plank, but it could not float them all, and the group disappeared underwater. Westbrook succeeded in grabbing on to a floating door or table and paddled around to the windward side of the steamer, where he joined Captain Miller at the paddlewheel. Westbrook described a quickly progressing fire, punctuated by several explosions and the sounds of gunshots. He estimated that not more than fifteen or twenty minutes passed between the outbreak of the fire and the emptying of the ship.[10]

The cause of the fire remained undetermined. Palace steamers seemed almost designed to burn. The wooden cabins, with their long salons and oil lamps, could be both fuel and furnace. The arson letter story was picked up by papers across the lakes, but no further details were ever forthcoming. Although not claiming arson, Captain Miller attributed the fire to human actions but was careful to deflect the blame away from the steamer and its crew:

Now, I am confident that the boat did not take fire from the machinery, nor from the boilers, as every portion of her fire-hold was fire proof. My opinion is, that the fire was caused by some combustible material stowed under the shafts, but the nature of which we were unable to tell, as packages frequently come so disguised that we cannot tell what they are; but it must have been something of that kind from the fact that it enveloped the boat in flames almost instantly; and when first discovered, it was impossible to subdue it.[11]

The *Niagara*'s walking beam remains well preserved, but its finer details are now obscured by invasive quagga mussels.
TAMARA THOMSEN, WHI IMAGE ID 140515

The *Niagara*'s paddlewheel assembly remain impressive despite damage from the fire, artifact hunters, and invasive mussels. TAMARA THOMSEN, WHI IMAGE ID 140398

The burning of the *Niagara* killed more than sixty people. The high death toll reflected two principal factors: the lack of established safety procedures on steamers and the architecture of the palace steamers themselves. The fire separated the officers from the stern of the ship, where three of the steamer's boats, including the largest, were located. This rendered it impossible for them to prevent the panicked passengers from taking matters into their own hands. Launching a boat from the deck of a ship requires skill and careful coordination. Terrified and untrained, the passengers succeeded in capsizing at least three of the *Niagara*'s five boats, while Congressman Macy's cowardice claimed a fourth.

A few days after the fire, eight of the *Niagara*'s surviving passengers issued a signed statement that declared the captain and officers blameless for the high loss of life and offering the following interpretation of events:

The *Niagara*'s anchor is on display at Harrington Beach State Park in Wisconsin, a lasting memorial to the steamer and her passengers and crew. FRIENDS OF HARRINGTON BEACH STATE PARK

> The overturning of the large boat in the stern quarter of the steamer was doubtless the occasion of the principal loss of life. Had the crew been there to attend to the lowering of the boat the accident might not have occurred, but it was not three minutes after the fire broke out before all communication was cut off between the bow and stern of the boat.[12]

Culpability for the disaster, these passengers felt, lay with unscrupulous merchants who heedlessly packed combustible materials like matches and fireworks among boxes and barrels of merchandise.[13] The lack of negligence attached to the owner, captain, and crew had important legal implications. In the absence of negligence, according to a law enacted by Congress in 1851, the victims of fires at sea were not entitled to any compensation for their losses.[14] For the purposes of this federal law, the Great Lakes were considered ocean and not inland waters.

The burning of the *Niagara* with its approximately sixty casualties was, after the *Phoenix* disaster, the second-deadliest steamboat fire to occur in Wisconsin waters. But fire was not the only threat passengers faced when traveling on Great Lakes steamers during the pioneering era of the wooden age. As the architecture of the powerfully built *Niagara* illustrates, shipbuilders such as Jacob Banta conceived of the Great Lakes of the mid-1840s as an especially dangerous place for large steamers. Banta designed the *Niagara* to take repeated beatings from storm, ice, collision, uncharted rocks, and frequent grounding in shallow water. Cutting through the enthusiastic hyperbole lavished on ships of the era, the *Niagara* was recognized as an uncommonly strong and safe steamer.

THE EVOLUTION OF PALACE STEAMERS

Dazzled by the size and luxury of the Great Lakes sidewheel steamers, travelers on the Great Lakes during the 1830s, 1840s, and 1850s took careful note of the maritime environment and of steamboat designs. These designs varied considerably and reflected changing economic conditions on the lakes and shifting fashions in American shipbuilding. During the 1830s and into the mid-1840s, Great Lakes passenger steamers possessed a regional distinctiveness clear to true experts such as the coastal engineer David Stevenson as well as to the observant general traveler. In the early 1840s, James Silk Buckingham saw the Great Lakes as a violent place and described its steamers as oceanic in size and in the staunchness of their construction. As for design, Thomas Horton James noticed that the lake steamers of 1845 were "built on a different plan, but not at all inferior, to the great steamers of New York."[15] One obvious difference during the 1830s and early 1840s was that passenger steamers in New York relied on external frames or trusses to achieve the required longitudinal hull strength. The trusses were standard in new Hudson River steamboats by 1832 and spread to steamboats built for coastal service on Long Island Sound and Lake Champlain during the mid-1830s.

Jacob Banta was part of an increasingly professionalized cadre of elite American shipbuilders. The social network of elite shipbuilders was small, tightly connected, and encompassed the Great Lakes and northeastern United States. Competition was rife, innovation the norm, design ideas freely debated, and promising ones quickly emulated. While Jacob Banta's design for the 825-ton, 225-foot long *Baltic*, completed in January 1847, was similar to the *Niagara*, his 906-ton, 245-foot long *Queen City*, completed in 1848, was substantially different. The *Queen City* was the first Banta steamer to employ the long, graceful external trusses that became known as bishop arches. The arches of the *Queen City* extended perhaps half the length of the ship and peaked well above the steamer's two-story cabin. All subsequent Bidwell & Banta sidewheel steamers employed some variant of external arches and associated trusses. Other Great Lakes shipbuilders adopted the Hudson River/Long Island Sound approach to their steamboat designs as well.[16] The use of the bishop arch spread and became the most visible regional characteristic of Great Lakes propeller-driven passenger and package freight steamers throughout the rest of the nineteenth century.

The shift from the homegrown designs of the early Great Lakes palace steamers such as the *Empire* and the *Niagara* in favor the Hudson River/Long Island Sound models stemmed from multiple causes. The Hudson River between New York and Albany was where Robert Fulton initiated the world's first successful commercial steamboat service. Its high traffic flows, periods of extreme competition, and early connection with railroads all helped to place an emphasis on the speed and efficiency of service and encouraged innovation. Although the lower Hudson River can be rough, the designers of its steamboats did not have to contend with violent seas regularly encountered on the coastal route between New York and New England or the Great Lakes. The need for northeastern coastal steamers to challenge rough weather was greatly reduced with the completion of railroads between Providence, Rhode Island, and Boston in 1835, and between Stonington, Connecticut, and Providence in 1837. Thereafter, the major passenger steamers were designed for the sheltered waters of Long Island Sound. The designs of the Long Island Sound steamers increasingly emphasized the economic-driven factors of size, speed, and comfort with less concern for the traditional dangers posed by the untamed ocean.

The later Bidwell & Banta palace steamers were faster and substantially larger than the *Niagara*. But where the *Niagara* had been built for a deep-water frontier understood as oceanic in its capacity for violence, the designs of the later steamers seem to suggest a more domesticated inland waterway. While diminished respect for the Great Lakes may account for some of the changes in steamer design, another likely factor is the technological hubris of American shipbuilders, whose short-term successes with giant sidewheel passenger steamboats bred overconfidence.

In 1861, English naval architect Norman Russell extolled the genius of US Hudson River/Long Island Sound steamer designs but also hinted at their potential for disaster in more hostile environments:

It is obvious, from the enormous dimensions of the ship, and the very slight thickness of materials used in the structure of the hull, that in so far as we have yet gone, such a deck, skin, and frame, are contemptibly flimsy, and according to our notions and practice quite incapable of carrying heavy boilers, powerful machinery, and a cargo necessarily uncertain in its nature, and unequal in its distribution. It is in the way in which he meets this, that the ability of the American builder is so conspicuously shown.[17]

Equally important to the enormous external bishop arches, the structural integrity of the Hudson River design also depended on a delicate series of internal masts and guy cables operated independently of one another to "keep some parts of the vessels down which have a tendency to rise, and keep others up which have a tendency to fall." Each mast and guy assembly had to be individually adjusted to "the exact strain shown by experience to be necessary." This level of engineering may have been fine in benign waters and in absence of trouble, but it left little room for even minor hull trauma.[18]

It is clear that the large Bidwell & Banta steamers that embraced the Hudson River/Long Island Sound designs were not as durable as the *Niagara*. Although the *Queen City* carried passengers for ten years before being laid up due to a financial downturn, the steamer sunk in shallow water at least twice after striking submerged objects. The *Queen City* lasted little more than a year after conversion to a lumber barge, when the hull broke up after striking a rock in 1863.

Comparison of the wrecks of the Bidwell & Banta steamers *Niagara*, launched in 1846, and *Lady Elgin*, built in 1851, offers material evidence of this shift from a Great Lakes design to the Hudson River/Long Island Sound design. Both steamers were designed by Jacob Banta and constructed at the Bidwell & Banta yard. They were nearly identical in size and cost and occupied the same economic niche, yet archaeological evidence reveals striking differences between the two. Substituting for the heavy internal arches and massive keelsons found in the *Niagara*, contemporary images and photographs of the wreck confirm the *Lady Elgin*'s reliance on bishop arches for longitudinal strength.

Compared with the *Niagara*, the hull itself was lightly constructed and contained perhaps half to three-quarters of the wood found on the older *Niagara*. Even in smaller details such as framing, the two steamers differed. The *Lady Elgin*'s frames were smaller and lacked the open-joint or split-frame arrangement found on the *Niagara*. From limited photographic evidence, it also appears that the *Lady Elgin* was less strongly fastened and probably did not have the heavy courses of thread-bolted thick strakes employed in the older steamer.

The wreck of the *Lady Elgin* on September 7, 1860, is one of the most famous wrecks in Milwaukee and Great Lakes history and tragically demonstrated the vulnerability of the northeastern steamer designs. The *Lady Elgin*'s port side perforated during a collision with a schooner about one-quarter its size, and the ship literally tore itself apart in about half an hour, drowning three hundred people.[19]

The wreck of the steamer *Keystone State*, a palace steamer built by Bidwell & Banta in 1849, also exhibits this shift to Long Island Sound design. Examination of an underwater video posted on the Internet by Michigan shipwreck hunter David Trotter, who located the steamer in Lake Huron in 2013, suggests that a similar structural failure occurred during the wreck event. While the heavily reinforced engine section and paddlewheels seem intact, the wreck's widely scattered condition suggests that the hull of the *Keystone State* failed and broke up during the storm that claimed it. Unlike the *Lady Elgin*, however, the *Keystone State* was not carrying passengers when it disappeared with twenty-three crew during a late November storm in 1861.

Other Great Lakes wrecks also suggest a shift to lighter construction techniques for Great Lakes steamboats. East Carolina University archaeologists investigating the wreck of the Canadian Great Lakes steamer *Maple Leaf*, built in 1851, also found the lighter hull construction. Launched the same year as the *Lady Elgin*, the *Maple Leaf*, like the American vessel, was

Sketch from the *New York Illustrated News* of the sinking of the sidewheel passenger ship *Lady Elgin* on September 8, 1860. A schooner struck the ship near Winnetka, Illinois, killing roughly three hundred people when the elegant hull broke apart. The depiction of the steamer is inaccurate but captures the human tragedy well. WHI IMAGE ID 6122

noted for speed but was better known for its economical construction than for its luxury.[20]

The burning of the *Niagara* was one of many significant disasters to occur in 1856. Just a few weeks later, the propeller steamer *Toledo* broke up during a storm on Lake Michigan while riding at anchor just off the pier at Port Washington, Wisconsin. Only three of the estimated fifty to sixty people on board survived. The 179-foot-long *Toledo*, built at Buffalo in 1854 by B. B. Jones and one of the largest of its class on the lakes, was completely destroyed, with cargo and wreckage strewn for miles along the shore.

Knowledgeable observers commented on the declining strength of Great Lakes vessels. Following the many shipping disasters of 1856, steamboat owner E. B. Ward analyzed Great Lakes navigation in a letter printed in the *U.S. Nautical Magazine and Naval Journal*. Ward blamed many problems on the rapid increase in the number of lake vessels and a decline the quality of their construction:

> The steam and sail vessels built during the past ten years have been generally too weak for their tonnage—too shallow in the hold—not much more than half fastened, and often overloaded. Hence the numerous losses by foundering of propellers and sail vessels, and the endless number of vessels springing a leak on the lake.[21]

The large number of serious accidents on the Great Lakes during the 1850s provoked significant responses by regional marine insurers who organized to promote safer navigation. Writing in 1857, the secretary of the Board of Marine Inspectors of the Association of Lake Underwriters discussed the dilemma facing insurers earlier in the decade:

> Lake Insurance was fast becoming a losing business, and as a great change had been made in the modeling, building, manning, equipping, loading, and navigating our lake vessels, it was reasonable to look upon this change as the main cause of this fearful increase of loss and its attendant consequences.
>
> The ship-builder bound down by a *low contract*, with abundance of competition about him, built to *suit*, as others built, and produced a vessel of *fine* model and *appearance*, of light draught and large carrying qualities, but possessing a very small amount of *wood*, *iron*, and *oakum*, for the size of the vessel.[22]

This Robert McGreevy illustration depicts the wreck of the former Bidwell & Banta–built steamer *Keystone State* as it looked when found in Lake Huron in 2013. ROBERT McGREEVY

The adoption of Long Island Sound–style steamer designs on the Great Lakes reflected regional pressures to achieve greater speed and capacity—pressures exacerbated by the need to deliver a large number of vessels at competitive prices. But even allowing for these factors, it appears that they were poorly suited for the rougher demands of Great Lakes service.

Expensive to build, run, and maintain, sidewheel palace steamers could thrive only while passenger volume and transportation costs remained high. The development of better passenger services on the railroads and the invention of screw-propelled freighters, which burned a quarter of the fuel and required smaller crews, spelled the end of the palace steamer era on the Great Lakes. By September 1857, some steamboat lines were losing $1,000 per day, and many of the newest and largest steamers were withdrawn from service forever.[23] Most of these vessels were scrapped or converted with their engines removed and their hulls recycled as barges, screw propellers, and dry docks.[24] Other sidewheel steamers would be built, but the elegant, finely lined steamers of the late 1840s and 1850s disappeared forever.

The archaeological record offers an interesting postscript to the palace steamer era in the wreck of the Canadian sidewheel steamer *Cumberland*, built in 1871. The archaeologists documented several architectural features strikingly similar to those found on the *Niagara*. Remarking on the lack of "the typical arched-truss support system," the archaeologists noted the presence of a wooden internal arch. Attached over the ceiling planking, the arch ran much of the length of the ship and rose nearly to deck level. The *Cumberland*, like the *Niagara*, had open-joint, or split, framing. On neither vessel did investigators find conclusive evidence of external bishops—arches or elaborate internal truss systems. Ironically, the *Cumberland* was powered by a recycled walking beam engine originally placed in service in 1846, the year of the *Niagara*'s launch.[25]

Why would a shipbuilder in 1871 resort to an 1840s design? Perhaps rather than being outdated, older designs remained appropriate for specific geographic and economic niches. On the Great Lakes in 1871, size and luxury lacked the status they had held during the palace steamer era. Competition was also less severe. The *Cumberland* was built for service on Lake Superior, the most oceanic of the lakes. Running well into the winter between Duluth, Minnesota, and Collingwood, Ontario, or Owen Sound, the *Cumberland* was expected to frequently encounter major storms and heavy ice. The builder's choice of the open-joint construction, heavy scantlings, and internal arching suggests that these features were associated with stout and resilient hulls. The *Cumberland*'s career, although not long, included many environment-related accidents. It ran aground on multiple occasions and was caught in an ice pack at least once. Fortunately, its hull, like that of the older but very similar *Niagara*, was very strong.[26]

Historical narrative and archaeological description and analysis do not always fold neatly together. Yet both approaches inform each other and it is with a blending of the two that a new picture of the *Niagara* and of the Great Lakes' palace steamer era has emerged. The charred and scattered wreck of the *Niagara* also offers a somber memorial to the dozens of men, women, and children whose lives were lost on a voyage to a new home in Wisconsin. The wreck also provides the general public, particularly the sports divers among them, with a recreational window on the past and a unique museum of antebellum steam technology and naval architecture.

On December 16, 1996, 150 years after the *Niagara*'s launching, the US secretary of the interior recognized its historical significance by placing the wreck on the National Register of Historic Places. Based on the work of the Wisconsin Historical Society, the University of Wisconsin Sea Grant Institute, and the Wisconsin Underwater Archaeological Association, the wreck was determined significant at the national level under each of the four National Register criteria and thus represents one of Wisconsin's most significant underwater treasures.

Ever the pioneering steamer, the *Niagara* became the first of what has become a statewide system of shipwreck mooring buoys. In June 1998, the Wisconsin Historical Society, in conjunction with the University of Wisconsin–Milwaukee, placed a special mooring over the wreck. Providing safer access for divers, the mooring has reduced the need for boats to anchor into the wreck and is helping to preserve the *Niagara* for generations to come.

Chapter Six

THE *LUMBERMAN*:
A SCHOONER OF THE
FOREST FRONTIER

The unexploited Great Lakes forests offered better entry-level business opportunities than steamboating. At the most primitive level, a healthy person with a sharp axe and requisite skill could transform a standing tree into a commodity with cash value—the stack of cordwood, the principal fuel for pioneer steamboats and the main heat source in the United States before the Civil War. A sawmill cost more than an axe, but the equipment was not beyond the reach of men with modest capital or solid credit. The sawmill was a necessity on the forest frontier and was always among the first industries established. In 1839, virtually every frontier community listed in the *Gazetteer of the State of Michigan* included at least one.[1]

The forests of the western Great Lakes began to generate serious wealth only when a large regional market developed. This began to take shape in the early 1840s and was centered in Chicago, located on the western shore of southern Lake Michigan. Chicago's strategic location between forest and the growing markets of the "Great West" led the city to emerge as the largest lumber market in the world by the 1850s. Ships were crucial to the Great Lakes lumber industry, and as production of lumber grew in western Michigan and eastern Wisconsin into the early 1890s, forest products became the most common type of cargo carried by ships on Lake Michigan. After the Civil War, the lumber trade made Chicago, by some accounts, the busiest port in the world in terms of vessel traffic. In 1872, at least 70 percent of all

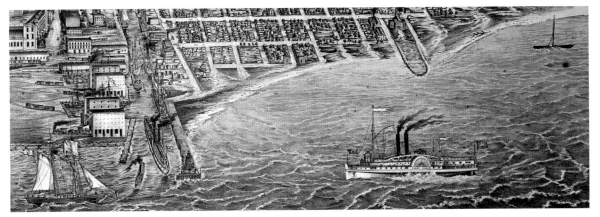

This detail of an 1857 bird's eye view captures a busy Chicago waterfront on a windy day. A steam tug tows a large barkentine, possibly the Bates-built *Mary Stockton*, into the narrow harbor entrance. Nearby is the *Lady Elgin*. Just off the beach to the north, a lonely mast protrudes from a shipwreck. LIBRARY OF CONGRESS GEOGRAPHY AND MAP DIVISION, 73693426

Chicago, the world's largest lumber market, at its peak in the early 1880s. *HARPER'S WEEKLY*, OCTOBER 20, 1883

commercial vessels landing cargo in Chicago—more than nine thousand—carried forest products.[2] Between 1872 and 1895, with the exception of three depression years, the volume of lumber transported by lake to Chicago exceeded 1 billion board feet every year. During the peak decade of the 1880s, ships delivered more than 15 billion board feet to the Windy City.

The staggering volume of the Great Lakes lumber trade during its peak and its subsequent decline are reflected in the composition of Wisconsin's historic shipwrecks. Forest products are by far the most common cargo that ships were carrying when they wrecked along the Lake Michigan coast and therefore dominate the underwater cultural landscape.[3] Forest and lake each offered aspiring entrepreneurs potential paths to wealth on the frontier. Those who combined the two had a clear advantage.

The inspiration and touchstone for this chapter is the *Lumberman*, a three-masted schooner built in Blendon Landing, Michigan. When launched in 1862, the *Lumberman* was arguably the most efficient lumber schooner on Lake Michigan. For the next thirty-one years, the ship carried lumber and related products from hinterland mills located along the coast and tributary waters of Lake Michigan to the Chicago market. The larger story, however, is about the place of ships in the interplay among the Great Lakes forests, waterways, and markets and in the personal and economic ambitions of two families of Michigan pioneers who sought to extract fortune from cutting trees.

Ambition on the Great Lakes Frontier

On a cold January day in 1868, the pioneers and elite of Grand Haven, Michigan, gathered to celebrate the life and mourn the passing of Reverend William Montague Ferry. Regarded as the city's founder and, for many, its spiritual leader, Ferry was also probably its wealthiest and most influential citizen. Among his extensive holdings was one-half interest in the *Lumberman*, a swift lumber schooner built at the behest of two of his sons in 1862. Ferry had first traveled west to the Michigan frontier not to acquire wealth but rather to spread the Protestant gospel and transform the region's "savages" into productive American citizens. In Ferry's view, the wealth that eventually followed was the just material rewards for a life of Christian service.

Born into a modest farming family in Granby, Massachusetts, in 1796, from an early age William Ferry was intent on becoming a clergyman. The path required expensive formal education that was beyond his family's means, so at age fifteen, he became a clerk in at a mercantile business in Utica, New York. For three years, he studied independently, saved money, and learned the ways of the business world—an education that would prove central to his later successes. He graduated from Union College in 1820 and, after additional study, was ordained by the Presbyterian Church in 1822.

Ferry secured a position with the new United Foreign Missionary Society, whose mission was to carry Protestant Christianity and the benefits of white civilization to the non-Christian world. The society dispatched Ferry to Mackinaw Island to study the feasibility of establishing a new mission. The trip introduced Ferry to the Great Lakes region's cultures and economy as well as to its ships and the men who sailed them. Ferry returned with an ambitious plan to establish a regional network of religious and cultural instruction for Indian peoples at Mackinaw. The Great Lakes fur trade, the US Army, and the work of God would all be based at Mackinaw Island. The society accepted Ferry's plan, and after an exciting journey that included passage on the pioneer steamer *Superior*, Reverend William Montague Ferry and his bride, Amanda Harwood Ferry, landed at Mackinac Island, Michigan, on Sunday morning, October 19, 1823.

Pioneer missionary and lumber baron Reverend William Montague Ferry, 1794–1868. LOUTITT PUBLIC LIBRARY

For the next ten years, William Ferry, his wife, and a growing staff of Protestant missionaries fought an uphill battle to establish and expand the mission and its school. The United Foreign Missionary Society folded in 1826, and the American Board of Commissioners for Foreign Missions assumed control. The culturally heavy-handed nature of Ferry's mission is clear in his first report to the new board. With regard to students attending the school, he noted that "children can as easily be obtained from the distance of several hundred miles, as from a much shorter distance." Once they were acquired and delivered to the mission, Reverend Ferry used the Northwest Territories indenture laws to bind the children to him and thus prevented their removal by "the caprice of their ignorant parents." He reported an enrollment of more than 160 students and accumulated physical assets worth $10,000 that included substantial buildings, a farm, and a new 18-ton schooner.[4]

Ferry's success was fleeting; his virulent anti-Catholicism and public battles with Catholic clergy alienated many people. Citing the expense and raising questions about the mission's effectiveness, the board dramatically downsized it in 1833, reassigning most of the staff to other missions, removing Reverend Ferry as the superintendent, and relieving him of all secular duties. Ferry's secular activities had been extensive and included investing in one and possibly two small trading schooners. Ferry had also been the center of a scandal involving his counseling of one of the female assistant missionaries—the charges were discounted by most but hotly circulated by his enemies. Citing poor health and the financial circumstances of his growing family, Ferry resigned his post with the board early in 1834.[5]

Along with a few enemies, Ferry had made many friends, most important among them Robert Stuart, the longtime manager of the American Fur Company's North Department. At the center of a widespread trading network, the wealthy Stuart had an exceptional knowledge of the economic, political, and cultural geography of the Great Lakes and extensive personal connections. Even before Ferry's resignation, Stuart had been guiding him in new directions. With Stuart's financial support and encouragement, Ferry circumnavigated Lake Michigan in a large fur trade canoe in 1833 and made his first visit to the American Fur Company's Grand River Post, managed by the colorful Rix Robinson and located at the confluence of the Thornapple and Grand Rivers and near the mouth of the Grand.

The mouth of the Grand River had obvious potential as a town site. Before 1831, the area had been part of the fur trade frontier, where property was held not by legal title but through possession and alliances between Native peoples and the American Fur Company. That year, federal government surveyors laid out the original boundaries for Ottawa County and several townships. The platting made it possible to acquire legal titles to frontier lands. On December 2, 1833, Rix Robinson, probably with Robert Stuart's backing, secured a patent for a large tract of land at the mouth of the Grand River. Ferry returned the following year, piloting a dugout canoe down the more than two-hundred-mile length of the Grand River, meeting up with Robinson on September 15, 1834.[6] Ferry moved decisively, and at the conclusion of this second visit, he and his brother-in-law Thomas White joined with Robinson and Stuart to form the Grand Haven Company.

At first glance, Ferry and Robinson had little in common. Robinson's Indian wives and mixed-race children, and his outstanding mastery of indigenous languages and place-based knowledge and skills, suggest a respect for the Native cultures and lifeways that Ferry had dedicated ten years of his life to destroying. However, seven years older than Ferry and also hailing from Massachusetts, Robinson possessed keen intelligence and was highly educated, having studied law in New York. Both men had a sharp eye for the new business opportunities that the Great Lakes frontier presented.[7] On Sunday, November 2, 1834, a little more than a month after chartering the Grand Haven Company, William and Amanda Ferry, their four young children, and all of their belongings arrived by schooner at the mouth of the Grand River. This is now considered the founding date for the city of Grand Haven.[8]

The mid-1830s were the halcyon days for land speculation along the Grand River, and the Grand Haven Company partners moved quickly. On April 14, 1835, Rix Robinson filed the original plat for the village of Grand

This 1874 bird's-eye view of Grand Haven, Michigan, shows busy lumber mills and the hazardous narrow entrance to the Grand River from Lake Michigan. LIBRARY OF CONGRESS, G4114.G4A3 1874 .R8

Haven. Two weeks later, William Ferry was appointed the village postmaster, a position he would retain for more than fifteen years. One reason for the haste was the pending dispossession of Ottawa and Chippewa from their land north of the Grand River. When this occurred (with Robinson's assistance) with the Treaty of Washington in 1836, the partners of the Grand Haven Company expected a land rush and quick profits. However, the Panic of 1837 ended the bonanza almost as soon as it had begun. The long depression that followed quashed the land rush and forced many speculators and recent immigrants to the Great Lakes frontier into bankruptcy.[9]

The village of Grand Haven and the Grand Haven Company survived, but their successes seem evenly mixed with failures, and the profits were minimal at best. Although his prospects for fast wealth imploded due to the business missteps of the Grand Haven Company, William Ferry continued to forge ahead. In 1840, despite the moribund economy, he opened a mercantile store at Grand Haven and built a lumber mill on the Grand River at the site of present-day Ferrysburg, Michigan. Ferry would maintain a diversified portfolio of business activities throughout the rest of his active life, but it was lumber that made and kept him and his family rich and powerful.

THE CHICAGO LUMBER TRADE

The output of Ferry's first mill and to whom he sold his products is unknown. What product he could not sell locally he likely shipped to Chicago.[10] In 1843 or 1844, he purchased the *Caroline*, an old 80-ton schooner. Although nearing the end of its useful life, the *Caroline* enhanced Ferry's position in the Chicago lumber market. With the economy looking bright at the beginning of the 1845 season, George Higginson, a Chicago lumber dealer, contracted with Ferry to deliver 1 million board feet of lumber at a price of seven dollars per thousand. Ferry could make such a contract only because he had his own schooner to deliver the product. Without the *Caroline*, he would have had to cope with the uncertain availability of ships and pay the going freight rate of $1.50 per thousand board feet. Owning the *Caroline* allowed Ferry to lock Higginson into a price that proved much higher than market average at Chicago for that season. Four decades later, Higginson was still grousing about the money he lost on the Ferry contract.[11] In this frontier period of the Lake Michigan lumber trade, even an old schooner could make all the difference, and it allowed Reverend Ferry to get in on the ground floor of what became the greatest lumber market in the world.

From its beginnings as a commercial port in the early 1830s, Chicago served a hinterland rich in soil and poor in trees. The completion of the Illinois and Michigan Canal in 1848 created a water route from Lake Michigan to the Mississippi River. Railroads followed on the heels of the canal, and by 1856, ten railroads connected the city to an immense and rapidly growing agricultural region. In practical terms, this meant that by the late 1840s, Chicago offered a nearly unlimited market for lumber, except during the periods of economic depression.

Between 1843, the year William Ferry's name first appears among the Chicago dealers, and 1848, the volume of lumber shipped to Chicago grew from 7.5 million to 60 million board feet.[12] In 1850 it topped 100 million board feet. The exponential growth in the lumber trade continued unabated through 1857, rising that year to 460 million board feet. Prices for lumber varied significantly but generally climbed in the early years, reaching as high as $17 per thousand board feet for clear pine lumber in 1855. At Chicago, however, high prices were not the norm. The plentiful and continuous supply of wood during the shipping season and the large number of producers serving the market institutionalized competition that kept Chicago prices low relative to other cities. The fact that a cargo of lumber could always be immediately sold for cash offset the relatively low Chicago prices.[13] Long-term success for lumber producers such as Ferry depended on the reliable production and sale of high volumes of product, and this required a secure supply of inexpensive timber, strategically located mills, and reliable transportation to the market at Chicago.[14]

Over the years, Reverend William Ferry and his sons operated or had investments in mills located along a variety of western Michigan's inland waters, including White Lake and the Muskegon, Grand, Pigeon, and Black Rivers. The inland waters brought timber from the forest to the mills. From the mill docks, the finished lumber and ancillary products such as shingles, lath, barrel staves, and railroad ties were loaded onto schooners and scows. The vessels then negotiated the often twisting and shallow waters down to the entrance of Lake Michigan. Wind, waves, and shallow waters with shifting channels often made the transit between inland rivers and the lake perilous. Even at the entrance to the Grand River, which has an area of deeper water that extends well inland, vessels seeking safe harbor at Grand Haven had to follow a constantly shifting eight- to ten-foot-deep sandy channel that was difficult, sometimes impossible, for a mariner to see. In light winds, the largest schooners and brigs of the 1850s and 1860s could safely land at Grand Haven, but in heavy seas, even a shallow draft faced destruction from a hard grounding.

On entering Lake Michigan, vessels had to cross more than one hundred miles of open and often rough water to the entrance of the Chicago harbor, where the convergence of large numbers of ships compounded the natural hazards of wind, wave, and shore. This need to navigate safely and profitably in shallow rivers and in the oceanic conditions found on Lake Michigan offered the principal challenges for the designers, builders, and operators of the lumber schooners. The particular circumstances of each river limited the maximum size of the vessel that could be used. Nonetheless, while Ferry's milling operations remained exclusively along the Grand River, large vessels could be used to carry the lumber to market. When operations spread out, as they did when Ferry and Sons built a mill on White Lake in Muskegon County in 1850, water depths diminished, increasing the need for shallow-draft vessels.

The Ferry lumber schooners changed in composition as the lumber trade and the family business expanded. Ferry's first lumber schooner, the *Caroline,* was built for the US Navy as a gunboat during the War of 1812 and had fought in the Battle of Lake Erie. The vessel's exact dimensions are unknown. At between 60 and 80 tons, the *Caroline* was a common size for an early trading schooner on the Great Lakes. Built as a highly maneuverable

At a lumber mill dock in Grand Haven, Michigan, in the early 1880s, men are rigging a new three-masted schooner, possibly the *Charles E. Wyman*. In the background, two schooners prepare to load cargo. ARCHIVES, CALVIN COLLEGE

gun platform, the hull was strongly built and likely fairly broad and shallow for its tonnage and thus a natural lumber carrier. Ferry squeezed that last few years of service from the *Caroline* before abandoning it in Spring Lake near one of the family's mills in 1848.

Ferry had his first new vessel built at Chicago in 1847. The two-masted schooner *Amanda Harwood* was 105.6 feet long, 23.2 feet wide, and 7.7 feet in depth. Measuring 170 tons (old measurement), the *Harwood* was a substantial vessel for the time. In 1848, Ferry established a shipyard upriver from Grand Haven at Mill Point (now Spring Lake) and hired shipwright John Conner to build the schooner *Telegraph*. Similar to the *Harwood* in size, the 179-ton schooner was 103.7 feet long, 24.5 feet wide, and 7.1 feet deep. The two schooners had maximum capacities of approximately 120,000 to 130,000 board feet of lumber. The *Telegraph*'s additional breadth and reduced depth should have allowed the schooner to draw somewhat less water when loaded than the *Harwood*.

The *Harwood*'s and *Telegraph*'s principal task was carrying lumber from mills on the Grand River to Chicago. However, they also had the important function of carrying freight from Chicago back to Grand Haven, principally to serve Reverend Ferry's mercantile store. Detailed descriptions of the vessels are lacking, but the schooners were probably of a more generic than specialized design. The *Telegraph* wrecked near Spring Lake in 1853, and it appears that the Ferrys sold the *Harwood* in 1854.

By the early 1850s, Ferry and Sons could sell lumber as fast as they could get it to Chicago. Because of the low profit margins and the constant need for cash to keep mills operating in the black, mill owners were under pressure to maintain low transportation costs and high-volume delivery through the season. For these reasons, subsequent lumber vessels built for or acquired by the Ferry family became more specialized.

The idea of specialized lumber-carrying ships goes back centuries. The shortage of shipbuilding timber in Britain and other European nations had required the importation of vast quantities of timber from North America and the Baltic and led to the construction of large oceangoing lumber vessels. By the 1820s, lumber carriers were common in New England, where the US lumber industry was first centered. What distinguished the lumber carriers from the other New England coasting schooners of the period was that they were configured to carry at least half of their cargo on deck. They also had reinforced deck stanchions and loading ports cut into the bow so that long boards and timber could be stowed below.[15] These ideas transferred naturally with the Atlantic migration to the Great Lakes of men involved in both the lumber and maritime trades. The lumber schooners that took shape in the 1850s and 1860s on Lake Michigan resembled their New England cousins but reflected the specific coastal geography and the faster pace and structure of the Chicago lumber market.

The first lumber boom overlapped with an exceptionally productive and creative period in the history of Great Lakes shipbuilding. The same talented shipbuilders revolutionizing the palace steamer also applied their efforts to sailing craft of various kinds, including those for the lumber trade. In 1853, Thomas W. Ferry contracted with Manitowoc shipbuilder and marine architect William Wallace Bates for the design of a new lumber schooner to be built at the Ferry and Sons shipyard on the Grand River. The twenty-six-year-old Bates had recently gained wide notice from the success of his "clipper" schooner, the *Challenge*. Bates based the design of the *Challenge* on principles published by John W. Griffiths in 1850. Built for the bargain price of $4,500, the *Challenge* was the fastest schooner on the lake, reaching speeds of up to fifteen miles an hour while in ballast and nine miles an hour when loaded. Seaworthy; highly maneuverable and equipped with a centerboard; and measuring 86 feet long, 22.4 feet in breadth, with a depth of hold of 6.5 feet, the *Challenge* could carry about 75,000 board feet of lumber while drawing about 5.5 feet of water. Empty, the schooner drew less than 3 feet. The *Challenge* wrecked in Wisconsin in 1910 and remains perhaps the most famous Great Lakes schooner ever built.[16]

Ferry came to Bates with very specific requirements for the schooner. It had to be fast in all winds, carry at least 90,000 board feet of lumber, have a depth of only a few feet when carrying a full cargo, and be capable of being loaded or unloaded in one day. For the Ferrys, Bates produced the design for the *Magic*, a centerboard schooner of modest length, just 93 feet, but with an expansive 26-foot breadth and a depth of hold of less than 7 feet. Reporting on the *Magic*'s launch, the *Grand River Times* noted that "her model is entirely new, at least in these parts, and her sailing capacities form a subject of much speculation among 'knowing ones'. . . . She is expressly designed for the lumber trade and is believed by persons who should be competent judges to be peculiarly well calculated for that purpose."[17] The initial results were encouraging, with the *Magic* stowing 49,000 board feet below deck and 47,000 board feet on deck on its maiden voyage. However, either the Ferrys or *Magic*'s builder had made a critical error in judgment.[18] Rather than following through on the Bates design, they kept the clipper hull form but eliminated a 20.5-foot-long centerboard in favor of a full keel. Under favorable conditions, the schooner flew through the water, "handled like a pilot boat," and could carry up to 110,000 board feet, the majority on deck. However, without the centerboard, the schooner lacked adequate lateral resistance to make good speed when sailing closer to the wind. Furthermore, the *Magic* proved cranky and vulnerable if poorly handled.

After a well-publicized capsizing in June 1855, Bates felt compelled to defend himself and the *Magic* in the influential *U.S. Monthly Nautical Magazine*, which he co-edited. He published an extensive analysis of the vessel's design that included a harsh indictment of Ferrys' failure to install the centerboard.[19] The decision not to install a centerboard was a significant departure from other Great Lakes schooners built during this period and remains something of a mystery. The deep keel would have increased the schooner's cargo capacity below deck by eliminating the centerboard trunk but also would have increased by several inches the amount of water the *Magic* drew when loaded. It may be that the Ferrys thought that the added keel would result in a faster or straighter-tracking vessel. Possibly, the English-born shipwright Fleming H. Revell, whom the Ferrys hired to oversee construction, had a hand in the changes. An experienced shipbuilder born in 1811, Revell had immigrated to Chicago in 1849 and may have applied his own ideas about Great Lakes ship design.[20] As a newcomer to the region, he may have seen the *Magic* as a step toward establishing his own reputation. Also in 1853, Revell built the schooner *Niagara* for the Ferrys. This vessel had virtually identical dimensions to *Magic* but lasted only two seasons before wrecking in Muskegon in 1855. The *Magic* survived twenty-six years of service, but the design failed to become the model lumber schooner that the Ferrys and Bates had anticipated.

At the beginning of 1857, the Ferrys purchased the schooner *Levant*. Built in Chicago in 1854, with a capacity of 110,000 board feet, the *Levant* was about 95 feet long, 23.25 wide, and had a surprisingly deep hold of 8.33 feet. The schooner had already proven itself carrying lumber from Muskegon to Chicago and probably had a reputation as a fine sailing vessel.

In addition to schooners, the lumber industry employed scow schooners. These were typically smaller, less expensive craft with schooner sailing rigs and much flatter bottoms. Under normal conditions, scow schooners did not typically sail as fast or true as regular lumber schooners; however, their low cost of construction and ability to navigate shallow water made them a popular choice for carrying lumber. They were also less likely to be damaged when run aground and easier to refloat.[21] The Ferrys owned several over the years and had at least two built. In 1854, they replaced the original schooner *Telegraph* with a scow schooner of the same name. At 79.5 feet long, 21.1 feet wide, and with a 6.9-foot depth of hold, the new ship carried 80,000 board feet of lumber. However, when loaded, the *Telegraph* drew only 3.5 feet of water and became the first sailing vessel to pass from White Lake into the White River to load lumber. In 1859, the Ferrys commissioned Fleming Revell to build the scow schooner *Storm* at the Ferry and Sons shipyard.

By 1856, the first phase of the lumber boom was at its peak, and the Ferry family's sixteen years in the lumber business had resulted in the ownership or control of multiple mills, a mercantile business with branches located near the mills, Grand Haven's first bank, a lumberyard and business located in a prime spot on the Chicago River, and the *Ottawa*, a new 315-ton passenger and freight steamer that ran between Grand Haven and Chicago. All of the Ferry sons became active in civic affairs, with second son Thomas W. Ferry elected to the Michigan legislature in 1850 and then serving a stint as an assistant collector of customs for the Port of Grand Haven. William Jr. had received multiple patents for milling machinery and was elected to the University of Michigan Board of Regents in 1856.

In 1857, the lumber boom came to an abrupt halt. That year, lumber deliveries to Chicago reached 459.6 million board feet. Although four and a half times the volume of 1850, it was just a whisker more than in 1856, and prices for lumber had plummeted. In 1858, Chicago lumber deliveries dropped below 300 million board feet. Although there was a slight recovery in production in 1859, the growing prospect of the Civil War tamped down the demand for lumber across the West and depressed demand and prices in 1860 and 1861. Tens of thousands of businesses, including many lumber companies, failed across the United States in the wake of the Panic of 1857, a financial crisis that led to an economic depression that lasted until the outbreak of the Civil War. Many of the pioneer lumbermen later remembered this period as the worst in the history of the lumber industry. By contrast, the Ferry family had amassed the liquid capital needed for lean times.[22] Good businessmen with diversified portfolios and well connected politically, the Ferry family had excellent credit and controlled every link in the chain of production, distribution, and financing of lumber and proved more than resilient.

BUILDING THE *LUMBERMAN*

Despite the depression, western Michigan's lumber frontier continued to attract new settlers. Among them was Allyne Cushing Litchfield, who in 1858 settled in Blendon Landing, Michigan, a heavily forested site on the Grand River several miles upstream from Grand Haven. At just twenty-three years old and recently married, Litchfield was an intense man determined to achieve both fortune and social position on the lumber frontier.

Born in Hingham, Massachusetts, in 1835, Allyne Litchfield was the son of highly respected East Boston shipwright Nichols Litchfield and the brother-in-law of Donald McKay, one of the most famous shipbuilders in the United States. McKay's clipper and packet ships embodied sophisticated design principles, and his yard was among the first to integrate steam saws into the shipbuilding process. Litchfield was raised and trained in the McKay shipyard, one of the most technologically advanced in the country. In short, he came west with advanced technical skills perfectly suited for wresting a fortune out of the Michigan forests. With financial backing from family and other partners, Litchfield established the mill at Blendon Landing in 1859. Capitalized at $40,000, equipped with forty steam-powered saws, and employing thirty-five hands, the mill could make up to 100,000 board feet of logs per day. Between July 1, 1859, and June 30, 1860, the mill produced some 4 million board feet of lumber.[23]

Litchfield was quickly counted among Ottawa County's elite. Registering as a Republican in Georgetown Township, Ottawa County, Michi-

Shipbuilder Allyne Cushing Litchfield (1835–1911), pictured in his officer's uniform before his capture by Confederate forces during a cavalry raid into Virginia in 1864. WHI IMAGE ID 139680

Donald McKay, world-famous builder of clipper ships and brother-in-law to Allyne Litchfield. THE METROPOLITAN MUSEUM OF ART

Thomas White Ferry (1827–1896), US politician and co-owner of the schooner *Lumberman*. LIBRARY OF CONGRESS BRADY-HANDY COLLECTION, LC-DIG-CWPBH-00494

gan, on April 4, 1859, he became an active member of the state party. This placed him in close company and political alignment with Edward, Noah, and Thomas Ferry, who were all active Republicans. Litchfield's religious outlook also aligned with the brand of evangelical Protestantism preached by Reverend Ferry and contributed to a growing social bond with Grand Haven's leading family.

The US Census schedule for 1860 reveals the wealth that was flowing into the hands of the Grand Valley elite. Reverend William Montague Ferry reported real estate worth $75,000. Sons Thomas, thirty-four, and Edward, twenty-three, remained a part of their father's household and did not report their property. William Ferry Jr., an engineer and manufacturer, reported $50,000 in property and $10,000 in personal estate. Lumberman Noah Ferry, twenty-eight, although unmarried, lived apart from his parents and reported $22,000 in property and $10,000 in personal estate. Newcomer Allyne C. Litchfield, aged twenty-five, also listed his occupation as lumberman and reported $3,000 in real estate and a dazzling $35,000 in personal estate.[24] Papers from the estates of Noah and Reverend William Ferry suggest, however, that they underreported their assets. Although he owned a valuable state-of-the-art mill, Litchfield, by contrast, was like most Michigan lumbermen—undercapitalized and vulnerable to any sudden economic misfortune.

Litchfield diversified his business by establishing a shipyard at the Blendon Landing mill site. The region's mix of hardwoods and white pine meant that Litchfield had access to prime but low-cost shipbuilding materials. The shipyard was located at an intersection between an unnamed tributary and the Grand River and connected to hardwood forests by the steam logging railway; it is difficult to envisage a better place to build a wooden ship in 1860. Expanding into shipbuilding at this time required little capital outlay, as building small to medium-sized sailing craft required little more than a suitable piece of ground adjacent to sufficiently deep and sheltered water, access to suitable building materials, a few hand tools, and sufficient skill. Litchfield had learned from acknowledged masters in the world of wooden shipbuilding and managed the most technologically advanced lumber mill in the region. It is possible, perhaps likely, that Litchfield expected to implement some of the more efficient milling technologies developed at the Donald McKay yard in East Boston. Given this package of skill, inexpensive resources, and economic opportunities in shipbuilding and ship repair, Litchfield's decision to establish his own shipyard appears sound.

Litchfield was more than a competent shipwright. He came out of the McKay shipyard with a highly developed understanding of modern ship design and construction methods. Litchfield's first schooner, the 246-ton *Major Anderson*, received unusually high praise, with the press describing the planking and finish work as the best yet seen on Lake Michigan. Litchfield laid down the *Anderson*'s keel in January 1861, and less than six months later, the schooner left Grand Haven bound for Boston, Massachusetts, carrying a full load of black walnut.[25] Three months later, Litchfield launched a slightly larger schooner that was purchased by Chicago interests.[26]

Sometime in 1861, however, Litchfield's good fortune left him. Litchfield had business connections with Augustus Paddock. Well-established in the area, Paddock had operated a steamboat on the Grand River as early as 1856 and had a part interest in the Litchfield mill, or the land that it occupied. The exact arrangements remain unclear, but things went bad, and Litchfield found himself in debt to Paddock and the Grand Rapids investors who were allied with him. The sum of the debt, undisclosed in Litchfield's letters, was large. Under increasing pressure, Allyne Litchfield struggled through 1861, managing to build the two large schooners but ending the year bitter and nearly bankrupt.

By contrast, the Ferry family continued to prosper. Reverend Ferry had largely withdrawn from active business, leaving the management of his affairs to his son Thomas. While his elder son William Jr. had joined the Union Army at the Civil War's outbreak, his younger sons Noah and Edward continued to aggressively pursue the lumber business, and prospects were looking good in 1862. By spring, Ferry and Sons was shipping out 200,000 board feet of lumber per week. Optimistic and possessing capital, Noah and Edward Ferry decided to build two new vessels, the new model steam tug *Come*, for use at their White River mill, and a schooner. It was not a scow schooner but a full-blown schooner specifically designed to deliver lumber from their mills to Chicago—the first such vessel for the family since the disappointing *Magic* and *Niagara* in 1853.

Why the brothers turned to Litchfield instead of building the vessel at the Ferry shipyard with Fleming Revell is unclear, but two possibilities come to mind. First, they surely recognized Litchfield as a shipbuilder of unusual ability. Second, they may have also been lending a helping hand to a worthy but struggling friend. In correspondence to Thomas W. Ferry written years later, Allyne Litchfield expressed thanks and acknowledged the Ferry brothers' decency to him during that difficult time. Whether it was to help, as good business, or both, it is clear that Noah and Edward Ferry expected something special from the schooner *Lumberman*.

The natural and economic conditions that had informed William Wallace Bates's design of the *Magic* in 1853 had only intensified by 1862. What the Ferrys needed was a relatively inexpensive-to-build shallow-draft schooner capable of making rapid voyages to Chicago carrying large cargoes of lumber. Quick loading and unloading was crucial—as were speed, seaworthiness, and vessel handling. The schooner that took shape at Blendon Landing was much different in design than the *Magic*. At 128 feet 5 inches in length, the *Lumberman*'s hull was more than 35 feet longer (37.6%), and yet had a maximum breadth of 23 feet 6 inches, some 2 feet 6 inches (9.1%) narrower. Whereas the *Magic* had a broad stern, the *Lumberman*'s hull tapered noticeably inward, narrowing to 18 feet 4 inches at the transom. The *Lumberman*'s depth of hold was at 7 feet, just 6 inches (9.3%) deeper than that of the *Magic* and no more than that of some scow schooners of the time.[27]

The length-to-breadth and length-to-depth ratios make these differences stand out, with the *Lumberman* at 5.4 to 1 and 18.2 to 1, respectively, and the *Magic* at 3.6 to 1 and 14.3 to 1. All things equal, this meant that the *Lumberman*'s hull was intrinsically faster but also potentially weaker than *Magic*'s and those of other schooners of similar proportion. Although not as dramatic a technical challenge as faced the designers of the *Niagara* and other palace steamers discussed in earlier chapters, the basic problem Litchfield faced was the same: how to build a vessel that was at once long, shallow in draft, fast, of high capacity, seaworthy, and strong. In the long, lean, twin-centerboard *Lumberman*, Litchfield met that challenge with flying colors.

Hundreds of schooners and scow schooners associated with the lumber trade wrecked in Wisconsin, but only four vessels originally built as schooners and one scow schooner have been archaeologically documented as having two centerboards. All of the schooners were built for the Lake Michigan lumber trade during the 1860s, with the *Lumberman* the oldest. Three of the schooners, the *Lumberman* (1862), the *Emeline* (lengthened and the second centerboard added in 1864), and the *Boaz* (1869), had very similar relative hull dimensions—long, narrow, and shallow. The fourth is the famous "Christmas tree ship," the *Rouse Simmons*, built in Milwaukee in 1868. The *Simmons* is a scant 3 feet shorter but a substantial 4 feet wider than the *Lumberman*. The *Simmons* depth of hold of 8 feet 1 inch is about a foot deeper than that of the *Lumberman*.[28]

Centerboard schooners came into use on the Great Lakes during the mid-1830s, and by the mid-1840s, they had become the norm for new construction and in many instances were installed in older vessels. Although they were associated with lumber schooners among Wisconsin's shipwrecks, the history of double centerboards on the Great Lakes has not been well documented. They did not originate on the Great Lakes. Probably the earliest known record of modern-style double centerboards is found in the plans for the *Union*, a fast shoal-draft schooner built in the United States in the 1820s and purchased by the Royal Navy for use against pirates and the slave trade in the Caribbean.[29]

The 1830s and 1840s were a time of much experimentation by American shipbuilders and the use of double centerboards spread to the Great Lakes by the late 1840s, where they found favor in larger vessels. The earliest Great Lakes reference to two centerboards found for this study is an unnamed 350-ton barkentine-rigged vessel built by D. Merrill and Company in Milwaukee in 1848. In 1852, the *Oswego Daily Times* reported the construction of a yet unnamed canal-sized barkentine at the innovative James A. Baker shipyard. Originally named the *Northern Light*, it was recently revealed to have two centerboards. It was re-rigged as a schooner and renamed *Montgomery* before wrecking near Sheboygan, Wisconsin, in 1890. In 1853, William Wallace Bates launched the much-commented-on 260-ton clipper schooner *Mary Stockton* in Manitowoc with two centerboards. In an article titled "Inland Navigation—Modeling Sail Vessels for the Lakes," written for the inaugural issue of the new *Monthly Nautical Magazine* the following year, Bates discussed the use of single and double centerboards and noted that "the adoption of two boards has followed partly as a necessity, and partly as an experiment, in adding the third mast." Bates cautioned readers that two boards "cost more than one board and are not so effective." Bates's comments clearly reflected his recent experience with the *Mary Stockton*, a vessel that newspaper reports alleged cost twice the normal price for a vessel of its tonnage. Notably, Bates described the appropriate hull for two centerboards as having "fine light ends, high, sharp bow, long midship body, with increasing sheer towards the head and stern, and accumulated strength amidships," a description largely in line with the *Lumberman* and its kin.[30]

Despite the expense and challenges, some builders of larger Great Lakes vessels continued to opt for two centerboards throughout the age of sail. For example, the 625-ton bark *Peshtigo*, launched by Wolf and Lawrence in Milwaukee in 1863; the 521-ton schooner *F. M. Knapp*, built by H. C. Pierson in Ferrysburg, Michigan, in 1867; the 744-ton bark *Annie Vought*, built by the Bailey Brothers Shipyard in Toledo, Ohio, also in 1867; and the Bailey Brothers' magnum opus, the 1,400-ton, five-masted schooner *David Dows*, launched in 1881, all carried two centerboards.[31] Although double centerboards appear to have been rare, their installation in the largest classes of sail vessels on the Great Lakes is easy to understand. Such craft were expensive already, and the larger hulls benefited from the increased lateral resistance the two centerboards provided.

The use of double centerboards in lumber schooners appears to have a more complicated history. The first mention of a Great Lakes lumber schooner with two centerboards is the *Shilleiah*, built by Frederick Nelson Jones in 1852, to sail between Saginaw and Buffalo.

The *Shilleiah*'s hull was very close in overall dimensions to that of the later *Lumberman* but was described as "flat-bottomed." How the schooner performed is unknown; other than a notice of its construction, the schooner seems to have slipped out of the historical record. Jones was well known as an innovative builder; however, the design does not appear to have caught on at the time.[32] Canal boats described as "surprisingly fast and stable" were equipped with sails and double centerboards and carried lumber and other cargo on the Cumberland and Oxford Canal from Sebago Lake to the harbor at Portland, Maine. The canal opened in 1832, but the double centerboards probably reflect a later period of use.[33]

Archaeological site plan of the *Lumberman*

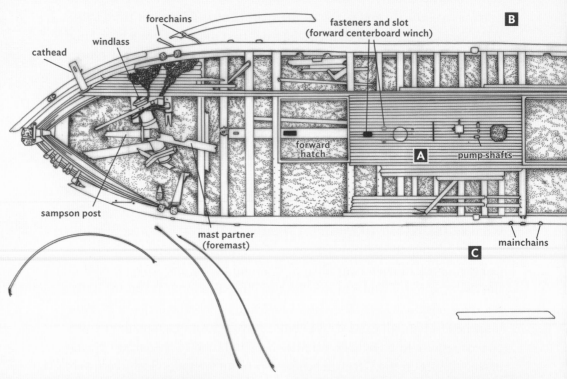

A. The intact centerboard trunk supports the main deck. The *Lumberman*'s two-centerboard design challenged the expertise of shipbuilders but resulted in a fast and profitable schooner. TAMARA THOMSEN, WHI IMAGE ID 140124

B. Divers survey the starboard side of the *Lumberman*. Missing deck planks reveal the strong deck beams that carried the heavy loads of milled lumber. TAMARA THOMSEN, WHI IMAGE ID 140125

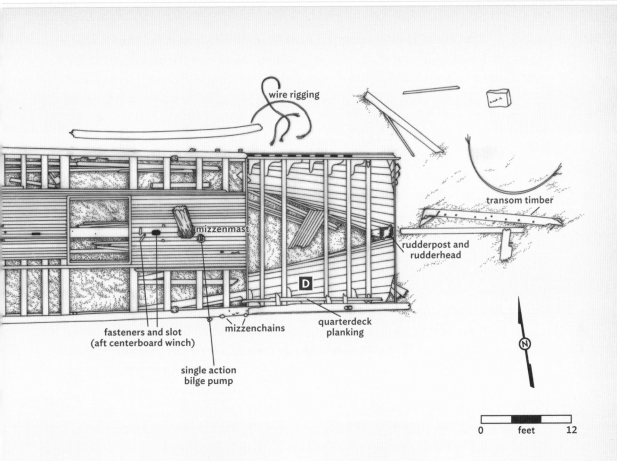

wire rigging

transom timber

mizzenmast

rudderpost and
rudderhead

D

fasteners and slot
(aft centerboard winch)

mizzenchains

quarterdeck
planking

single action
bilge pump

N

0   feet   12

C. Lumber ports to speed transfer of cargo were a standard feature on specialized lumber vessels such as the *Lumberman*. TAMARA THOMSEN, WHI IMAGE ID 140126

D. This view of the port side of the *Lumberman* was shot from near the stern and looks toward the bow. The intact beams in the foreground carried the weight of the cabin and are of smaller dimensions than the heavy beams that supported the main deck. The broken stump of the mizzenmast is visible in the background. TAMARA THOMSEN, WHI IMAGE ID 141895

The use of even one centerboard brought challenges for shipbuilders and owners. Encased in a watertight box called a centerboard trunk, the centerboard potentially compromised the water integrity and longitudinal strength of the hull. During the 1840s and 1850s, shipbuilders on the lakes often located the centerboard assembly just offset from the centerline keelson. According to Bates, the arrangement made it easier to mast the vessels and the unbroken centerline keelson/keel made for a hull stronger. By the early 1850s, some shipbuilders, including Bates, began producing schooners with the centerboards located atop the centerline keelsons in smaller vessels, such as the *Challenge*. This soon became the standard for all sizes of crafts. Responding to high levels of losses, the Board of Marine Inspectors of the Association of Lake Underwriters (the regulatory arm of the lake marine insurance business) published its first construction recommendations for new vessels in 1856. The board recommended centerline boards in new construction. The published insurance rates for 1856 also indicate that vessels in the lumber trade were considered a higher risk. While the owner of a new vessel of 200 tons would pay a premium of 7 percent of the hull value insured, lumber vessels loading from western Lake Michigan ports paid 8 percent, and those from eastern Lake Michigan ports such as Grand Haven paid 9 percent.[34]

By 1862, shipbuilders had addressed any structural weaknesses associated with single-centerboard vessels. But two centerboards seriously compounded the problem. In retrospect, given the combination of the *Lumberman*'s long, narrow, and shallow dimensions, it seems that building sufficient hull strength into the schooner with one centerboard would have been challenging. Adding two 20-foot-long centerboards in a schooner intended to carry heavy cargo must have required building in a great deal of the "accumulated strength amidships" Bates specified in his 1854 discussion of double centerboards. Indeed, the *Lumberman*'s combination of three masts with a hull characterized by "fine light ends, high, sharp bow, long midship body, with increasing sheer towards the head and stern" conforms with the ideal characteristics for double-centerboard sailing vessels Bates had laid out.[35]

That Litchfield met the challenge is obvious, as the *Lumberman*'s career spanned three decades. However, the surviving portions of the *Lumberman*'s sunken hull described in archaeological reports are suggestive of but inconclusive on what made the *Lumberman* so durable. Multiple rebuilds over the course of the schooner's career (1871, 1878, 1879, 1889) may have introduced substantial alterations to the ship's architecture. As the ship remained in the same trade as a highly successful carrier and the rebuilds were not due to catastrophic events, any structural changes were probably minor. However, renewing the decks, bulwarks, and exposed area of exterior planking would have been necessary, and replacements may have introduced some significant changes.

To add strength, the builders of larger schooners on the lakes frequently installed wooden arches in the form of a second layer of edge-joined (iron fastening running vertically down through the boards) ceiling planking. The *Lumberman*'s shallow draft precluded this type of reinforcement. Underwater archaeologists did report the installation of extra-thick exterior planking (thick strakes) at the turn of the bilge—the vulnerable place where the sides and bottom of the hull join. While the bulk of the exterior planking recorded varies from 1¾ to 2 inches in thickness, the thick strakes are estimated at 3½ inches by 3½ to 4 inches in width. Great Lakes builders often added considerable strength to their hulls with the heavy ceiling planking. On the *Lumberman*, the 2½-inch ceiling planks seem substantial for a schooner of its tonnage. The planks varied in width between 10 and 11 inches, even more important to adding hull strength than the thickness. While not certain, the ceiling planks were probably edge-joined to create a solid wall-like structure, at least above the turn of the bilge. The exterior and ceiling planks were fastened to the *Lumberman*'s frames with half-inch square iron nails or spikes. The schooner's rather flat sides and the wide ceiling planking, in combination with the thick strakes and solid exterior planking, should have imparted considerable longitudinal strength or stiffness to the hull. However, this would have depended on strong fastenings. These would gradually loosen over time through corrosion, wood decay, and the working of heavy seas. The need to renew fastenings may help explain much of what occurred during *Lumberman*'s multiple rebuilds.

Lumber schooners carried more than half their cargo on deck. The *Lumberman*'s typical deck cargo weighed somewhere between 200,000 and 250,000 pounds. The decks of traditional schooners not built for the lumber trade were supported by beams that rested on L-shaped timbers called *hanging knees*. However, traditional hanging knees reduced the capacity to carry long boards below deck, and instead of knees, the *Lumberman*'s deck support came from beams that were tied to a long timber that ran along the inside of the hull, called a *shelf piece*, and posts called *stanchions* that extended from on top of the keelson to the deck beams. The shelf piece and stanchions transferred the weight carried by the beams to the ship's frames and keelson, the strongest parts of the hull.

The deck beams on the *Lumberman* are heavy, at 11 inches thick (molded) and between 9 and 10 inches wide (sided). The beams on each side of the three cargo hatches are wider, at 13 inches, but of the same thickness. The spacing between is close. In the roughly 35 feet separating the two center-boards, seven beams supported by stanchion posts support the deck. Varying from 5 inches by 5 inches to 5 inches by 7 inches in dimension, the stanchions seem a bit small compared to the deck beams. However, the joints between stanchion and beam are strongly fastened with iron hasps. Heavy iron tie rods are used to connect the deck beams with the keelson at the back end of both centerboard trunks. Collectively, the midship section of the *Lumberman* appears strongly put together but exhibits more finesse in architecture than inherent strength from heavy materials. With the possible exception of the spacing and size of the deck beams, the hull's planking and timbers are substantial but not unusually large in their dimensions. This make sense, as Litchfield would have wanted to build the lightest hull possible consistent with the physical demands of the lumber trade, and every extra pound built into the structure pushed the hull just a little deeper into the water, reducing speed, shallow water navigation ability, and cargo capacity.

Neither Litchfield nor the Ferry brothers left records explaining why they opted to build a long and narrow schooner with two centerboards and three masts, and the matter is not discussed in regional newspapers. Yet by the end of the decade, the *Lumberman* had inspired a number of similar lumber schooners, including at least two, the *Emeline* and the *Major N. H. Ferry*, that were owned by the Ferry family.

On July 16, 1862, the *Grand Haven News* took notice of the nearly finished *Lumberman*: "As her name, the *Lumberman*, would suggest, she is built with special reference to the lumber trade, and all her planning of hold and deck evidences the mind of a practical man as fully as her model and fastenings the skill of a master-builder." The paper took note of the two centerboards and reported the vessel's cost at around $10,000. Upon its completion, the newspaper was more lavish in its praise, stating that "the *Lumberman* . . . is acknowledged to be one of the finest crafts ever put in commission for the lumber trade." Allyne Litchfield filed the master carpenter's certificate for the *Lumberman* on August 2, 1862. The vessel's first enrollment was secured by Noah H. and Edward P. Ferry on August 8. The original measurements recorded for the *Lumberman* were 128 feet 5 inches long, 23 feet 6 inches in breadth, and 7 feet 2¾ inches in depth, with an "old measurement" tonnage just over 204 tons. The Ferry brothers placed veteran mariner Michael Connell in command. Connell had previously served

the Ferry family as captain of the schooner *Levant*. On August 14, 1862, the *Lumberman* delivered its first cargo, 150,000 board feet of pine lumber, to the Ferry lumberyard at Chicago.[36]

THE *LUMBERMAN*'S CAREER

Newspaper coverage of the *Lumberman*'s long operational history is extensive. For some periods, it is complete and provides us reliable data to make some assessments of the vessel's qualities.[37] Two related facts particularly stand out about the *Lumberman*: the rapid turnaround when in port and the large number of trips it made each season. When contracting for the design of *Magic*, the Ferrys specified a schooner that could be loaded or unloaded in one day. Litchfield achieved this important benchmark in the much larger *Lumberman*. The schooner frequently entered the Chicago River with a full cargo of 140,000 to 150,000 board feet of lumber and related products, unloaded, and began its return voyage on the same day. This represents remarkable efficiency. When the lumber industry was well established on the Pacific coast at the beginning of the twentieth century, sailing vessels remained the most economical transport for coastal lumbering operations. According to a detailed account printed in the *Marine Review*, the usual rate of loading or discharging cargo from a sailing vessel on the Pacific coast in

1909 was 75,000 board feet per day—one-half the speed achieved with the *Lumberman* and similar schooners.[38]

Between June 17 and July 2, 1872, the *Lumberman* delivered five cargoes totaling 665,000 board feet of Michigan lumber to Chicago. While this may have been an especially busy period, it illustrates the remarkable efficiency achieved with these specialized schooners. Under normal conditions, the *Lumberman* apparently would deliver a typical cargo to Chicago every four to five days. In 1870, newspapers recorded twenty-nine arrivals at Chicago for the *Lumberman*. Analysis of gaps in the dates suggests that the ship made at least ten to twelve deliveries that went unrecorded. Based on a rough average of 140,000 board feet or the equivalent per trip and an estimated forty voyages, the *Lumberman* delivered 5.6 million board feet of lumber to Chicago in 1870. The information for the 1870 season is especially complete, but data available from other years suggest that this was a normal season for the *Lumberman*.

The double centerboards contributed to the *Lumberman*'s performance in at least three ways: overall speed, maneuverability, and stability. The schooner's long, narrow, shallow, and carefully modeled hull was naturally fast through the water. The basic shape, however, offered little lateral resistance when under sail. The two 20-foot-long centerboards provided that resistance, but in contrast to a single-centerboard schooner, the resistance was distributed farther forward and farther aft along the hull. The location of the two boards in combination with the three-masted topsail schooner rig allowed the vessel to track straighter through the water than a typical one centerboard schooner, adding to its overall speed between the Ferry lumber mills and the Chicago market.

Sailing safely and well with a large cargo of lumber piled high on deck required a lot of stability. The shallow and narrow hull, however, offered little natural stability. As the schooner cut through the water under sail, the two centerboards bit into the water and created forces that helped to keep the hull from heeling (leaning) over too far. The two boards could also be adjusted to help "trim" or balance the schooner, a critical feature for sailing fast and safely with the heavy deck loads. The balance achieved by the three-masted topsail schooner sailing rig and the two centerboards provided the *Lumberman* with excellent stability while under way. In terms of handling, with the skillful use of the double centerboard, the schooner could be made to turn faster by retracting one board and using the second as a virtual pivot point. The additional handling capabilities contributed to the success of double-centerboard racing yachts and would have done the same in the lumber trade.

The *Lumberman* illustrated all of the positive aspects of the narrow double-centerboard three-masted schooner needed for the Chicago lumber trade. However, such a vessel was not only more complicated to build, but also required a higher degree of seamanship by captain and crew. Michael Connell was replaced as the *Lumberman*'s captain by Richard Williams in 1865. Williams would remain in command until Edward and Thomas Ferry sold the *Lumberman* in 1883. This consistency of command and obvious mastery of the vessel's strengths and vulnerabilities contributed to its success and longevity.

At the peak of the trade, the pace of maritime life on Lake Michigan's lumber schooners was probably the fastest in the world. The constant turn-arounds kept the men extremely busy. On the mill side of the journey, the crew had the backbreaking chore of loading the cargo—this work had to be done quickly but also correctly to ensure proper trim and to reduce the chances of its shifting. On the lake, racing among the schooners was reportedly common. For independent vessels that might have to wait in line for days to get to the dock to unload, the racing might have real economic significance. Owned by a family of lumber dealers as well as mill owners, the *Lumberman* did not have to wait in line, but pride and the pressure to maximize the number of trips completed during a season provided plenty of reason to show speed.

The need for speed, the pressures of competition, and high volumes of traffic combined with the natural forces of Lake Michigan to make the lumber trade a dangerous business. Like all schooners of long service, the *Lumberman* was involved in a number of minor as well as more serious incidents.

With more than twelve thousand vessels landing at Chicago in the 1870s, most of them in the lumber trade, collisions were, not surprisingly, the single most common type of accident sustained by the *Lumberman*. On October 8, 1869, the large barkentine *Hungarian* collided with the *Lumberman* outside of the Chicago harbor, staving in the stern and destroying its small boat. The resilient schooner was quickly repaired and delivered another cargo less than a week later. In June 1874, the *Lumberman* sustained minor damage in collision with the schooner *Pilgrim*. In October 1877, while under tow in the Chicago harbor, the *Lumberman* crashed into the steamer *Egyptian*, suffering slight damage. The following year, the *Lumberman* collided with the tug *Rebel*. On a dark November night in 1881, a schooner tentatively identified as the *Union* collided with the *Lumberman* at high speed and then sailed off. There were many other such incidents.[39]

Sudden bad weather brought potentially fatal dangers. While en route to Chicago from Grand Haven with a cargo of lumber on November 30, 1867, the schooner got caught in a heavy gale and was pounded so heavily in the waves that both centerboards broke off. That the vessel made it back to port unassisted and with its cargo intact attests to both the quality of the schooner and the seamanship of Captain Williams.[40] In October 1877, the *Lumberman* lost part of its deckload in a storm. In November 1879, the *Lumberman* and its near twin, the *Major N. H. Ferry*, successfully negotiated a forty-mile-per-hour gale to make safe harbor at Chicago. In September 1880, a strong northwesterly blow carried away sails and broke the schooner's main boom and main gaff. On July 2, 1882, a rare summer northeasterly squall drove several loaded schooners, including the *Lumberman*, ashore south of the Chicago harbor. The vessel was undamaged. Not all dangers came on the lake; in September 1882, thieves at Chicago cleaned out the belongings of the *Lumberman*'s sailors, scouring the forecastle.[41]

The most unusual accident that the *Lumberman* incurred took place on a hot night in late August 1883 when a lightning bolt struck the mizzen topmast, causing it to explode. The bolt traveled through the wet mast timbers, shattering the binnacle box, temporarily paralyzing the wheelsman, and burning one side of Captain Williams's face. The *Lumberman*'s mate described the strike as producing thousands of sparks, similar to a grenade explosion. The damage, however, did not prove severe, and the vessel again made it safely back to Chicago.[42]

The accidents and dangers faced by the *Lumberman* and its crew during the twenty-one years that the Ferry family owned the schooner were typical of the Great Lakes. Indeed, the *Lumberman* seems to have had far fewer and less serious accidents than most schooners during the period. The vessel was clearly well handled, profitable, and well maintained, with routine annual upkeep and multiple major rebuilds when needed.

The *Lumberman* should be seen as an important component of a larger technological and economic system that transformed frontier forests into commodities at a rate previously unseen in human history. The system transformed trees into cash. Although the individual profit margins were typically low, the volume of lumber products could make up for it. The Ferry family controlled every facet of the lumber industry, from cutting to selling. In 1874, Ferry's White Lake mill alone produced an average of 75,000 board feet of lumber and 50,000 shingles per day. Two days of lumber production would have been a full load for the *Lumberman*, and the Ferrys had acquired a small fleet to keep the product flowing, including two other double-centerboard schooners inspired by the success of the *Lumberman*.

In the Wake of the *Lumberman*

The launching of the *Lumberman* marked the end of Allyne Litchfield's shipbuilding career. Facing bankruptcy and energized by the outbreak of the Civil War, Litchfield closed up his mill and shipyard. Less than two weeks after completing the *Lumberman* in 1862, he enlisted in the Fifth Michigan Volunteer Cavalry, receiving a captain's commission. Fighting in illustrious company, including that of General George Armstrong Custer, Litchfield received promotions to major and then lieutenant colonel. On March 1, 1864, he was captured in the disastrous Kilpatrick-Dahlgren Raid on Richmond, an ill-advised attempt to free Union prisoners of war.[43] Litchfield endured solitary confinement and torture and was at one point condemned to death.[44] He spent the next year in a series of prison camps. During his confinement, Litchfield gained promotion to full colonel. As recognition for his battlefield gallantry, Litchfield received a brevet promotion to brigadier general. The brevet promotion brought no salary, pension, or high command but did provide Litchfield with the socially useful postwar title of general.

During the war, Litchfield lost his mill to Augustus Paddock and his cronies, but he had the satisfaction of knowing the mill burned to the ground shortly afterward. In May 1867, Allyne Litchfield's brother Lawrence purchased the former Union gunboat *Trefoil* for the bargain price of $11,500. Built in the closing days of the war by brother-in-law Donald McKay, the quick and seaworthy vessel proved perfect for the Great Lakes passenger and package freight business. Equipped with new cabins and renamed the *General H. E. Paine*, the steamer went into freshwater service with Litchfield serving as its clerk. Litchfield settled his family in Northport, Michigan, where his position as steamboat clerk brought in a modest living. In the 1870 US Census, Litchfield reported owning personal property valued at $570, but no real estate. This was a severe reversal for a man who had once owned the finest lumber mill on the Grand River.[45] With the patronage of the newly elected US Senator Thomas W. Ferry, on May 22, 1871, Litchfield received an appointment as US Consul General to British East India in Calcutta, a diplomatic and commercial post he retained for the next ten years. But Litchfield's wartime trauma deeply affected his emotional and physical health. After returning to the United States in 1881, he held a series of short-term jobs before settling near his son, a successful physician in Pittsburgh, Pennsylvania, where he died in 1911.

Despite the delivery of the new schooner and a booming business, Noah Ferry, like his friend Allyne Cushing Litchfield, was eager to serve the Union cause. Leaving the business in his brother Edward's hands, Noah Ferry accepted a captain's commission in the Fifth Michigan Volunteer Cavalry and went to war. Battlefield fortunes did not favor Noah Ferry. A new vessel enrollment for the *Lumberman* issued on August 13, 1863, silently removed Noah Ferry's name as the schooner's co-owner. He had been killed fighting alongside Litchfield at Gettysburg.[46] With Noah's death, full ownership of the *Lumberman* passed to Edward P. Ferry. In August 1865, for unknown reasons, Edward transferred the vessel to his father, who retained possession until his own passing on December 30, 1868. On March 26, 1869, one-half ownership passed back to Edward, with the balance belonging to his brother Thomas W. Ferry. The two men retained joint ownership of the schooner for the next fifteen years.[47]

As the *Lumberman* safely carried their lumber, Edward and Thomas Ferry's economic and political fortunes continued to grow. Elected to the US House of Representatives in 1864, Thomas Ferry proved an active congressman, especially in areas of finance. In 1871, Michigan elected him to the US Senate. In a time with less sensitivity to conflicts of interest, Senator Ferry worked tirelessly to protect the interests of midwestern lumbermen and lake interests. On March 9, 1875, Ferry was elected president pro tempore of the Senate—a position he held for the next four years. With the death of US vice president Henry Wilson on November 22, 1875, Ferry became acting vice president. For more than a year, the *Lumberman*'s co-owner was a heartbeat away from the presidency. In that role he presided over the commission that settled the controversial presidential election of 1876.[48]

As Thomas Ferry helped govern the nation during trying times, his brother Edward focused on the business and did well for many years. Harsh changes, however, came in the winter of 1882–1883. The Ferry brothers had speculated heavily in Utah silver mining. In February, the firm declared bankruptcy, and its assets, including the *Lumberman*, were sold. During the same period, Thomas lost a bitter battle to retain his Senate seat and had become the target of public scandal. He fled to Europe for three years before returning to Grand Haven, where he died in 1896.[49] Edward Ferry managed to recover his fortunes in the West. Settling in Park City, Utah, he was elected to the Utah state legislature. Reverend William Ferry had named

Edward executor of his estate. In 1903, Edward's sister Amanda Hall filed a lawsuit to have him removed as the executor of their father's estate. After numerous appeals, in 1913 the US Supreme Court ruled against Edward Ferry and determined that he owed the estate nearly a million dollars. By this time Ferry had been institutionalized as mentally incompetent. His son Edward Jr., a lawyer who had represented him in the case, became despondent and committed suicide. Two years later, Edward's elder son, William Montague Ferry III, was elected mayor of Salt Lake City, Utah.[50]

Reverend Ferry's eldest son, William Montague Ferry Jr., had no direct interest in the *Lumberman*, but he was involved with his brothers in milling businesses. The one Democrat in the family, William was elected mayor of Grand Rapids, Michigan, before leaving to join Edward in Utah in 1878. In 1904, he ran unsuccessfully for governor and died the following year.

THE *LUMBERMAN*'S LATER YEARS

In March 1883, Isaac Lombard of Chicago purchased the *Lumberman*. The vessel continued in the lumber trade under the supervision of Captain Williams. Thirteen months later, the partnership of Harry Smith and John Lang purchased the vessel, also, presumably, for the lumber trade. These men in turn sold the vessel to V. Mashek of Kewaunee, Wisconsin, and Captain Arnt T. Hansen of Chicago for $2,500. The Mashek and Hansen partnership changed its port of registry from Chicago to Milwaukee and continued as the *Lumberman*'s owner until its loss in 1893.[51]

Under Mashek and Hansen's joint management, the *Lumberman* carried lumber, bark, shingles, and other forest products from small Wisconsin lumber ports such as Lily Bay and Whitefish Bay to Chicago. Occasional notices of the *Lumberman* appear in the *Door County Advocate*, a small newspaper that kept close track of Lake Michigan ships and shipping, as well as in the Chicago newspapers. Because they carried wood, a buoyant cargo not easily damaged by water, the old schooners used to carry lumber often received only minimal maintenance in their later years. The *Lumberman*, however, was never considered a used-up or second-class vessel. Its later owners continued to keep the schooner well maintained. For example, in June 1890, the *Manitowoc Pilot* reported that Captain Hansen was conducting "thorough repairs" of the *Lumberman*.[52]

On the afternoon of April 6, 1893, Captain Orin Vose should have been a happy man. He had just embarked on his first trip as master of the veteran schooner the *Lumberman*. Early that morning, Vose took his vessel out of Chicago to begin a quick journey to Kewaunee, Wisconsin. His job was to secure a load of forest products for the Chicago markets. Running light, the long shallow schooner made good time on this first trip of the season. Unknown to Vose, however, who was resting in his cabin, a vicious storm system from the southwest was racing across Lake Michigan. In an age before radio, the seven-man crew could not know that the storm system had already caused terrific damage in several Illinois communities.

The southwesterly winds hit the *Lumberman* hard, knocking it over onto its starboard side. The schooner quickly filled with water and sank to the bottom of Lake Michigan. Fortunately, the water was not deep—just sixty feet—and the vessel righted itself as it went down. Although Captain Vose was trapped by the rigging, pulled underwater, and nearly drowned, the entire crew ultimately scrambled to safety in the tall topmasts that protruded from the lake. Three hours later, the steamboat *Menominee* rescued them from their damp perch. The owners wanted to refloat the schooner, but despite the intentions of an enterprising Racine tugboat operator, Edward Gillen, the *Lumberman* was never fully salvaged. At end of June 1893, Gillen removed the *Lumberman*'s spars but left the hull undisturbed. By late August, the *Lumberman* had become one more of thousands of abandoned shipwrecks scattered across the bottom of the Great Lakes.[53]

The commentary from the maritime community at the time of the *Lumberman*'s loss is telling. The vessel was considered to be in first-class condition, a fact reflected in its A2½ insurance rating, which was unusually high for a thirty-year-old schooner. The *Racine Journal Times* reprinted a recent column from the *Northwestern Lumberman* that commented on the *Lumberman* and its younger but nearly identical sister, the *Major N. H. Ferry*. While the *N. H. Ferry* was "strong headed and rebellious," the *Lumberman* "always acted like a thoroughbred lady, and so much confidence did her owners have in her that the lumber she brought over came uninsured."[54]

How many double-centerboard schooners were built for the lumber trade on Lake Michigan or wrecked in Wisconsin remains unknown, but those that have been identified exhibit remarkable longevity as a class. The *Lumberman* is the oldest, followed by the *Emeline*, which had been purchased and transformed into a smaller version of the *Lumberman* by the Ferrys in 1864. The *Emeline* wrecked in 1896, and its remains lie in shallow water in Baileys Harbor, Wisconsin. With the departure of Litchfield, H. C. Pierson became the Ferrys' builder of choice, and he produced several schooners at Ferrysburg, Michigan. Despite a reputation for ill luck, the *Major N. H. Ferry*, which was built for the Ferrys in 1868, continued in service until it sank in a collision on Lake Erie in 1913. The *R. P. Mason*, built by Pierson in 1867, had very similar dimensions and almost certainly had two centerboards. The *Mason* was eventually cut down into a barge but survived until 1917 before burning on Lake Michigan in Wisconsin waters. In 1867, Pierson also built the much larger *F. M. Knapp.* Designed for the grain trade, the schooner also carried lumber and was converted to a tow barge and finally abandoned in 1930. The *Rouse Simmons*, the famed Christmas tree ship, built on a slightly different model by Allen, McClellend and Company in Milwaukee, continued sailing until its loss northeast of Rawley Point, Wisconsin, in November 1912. Although rough in its construction, the *Boaz*, built by Amos Stokes in Sheboygan in 1869, is clearly modeled on the *Lumberman* style. The *Boaz* remained in service until 1900. The vessel's wreckage is easily accessible in shallow water in North Bay, Wisconsin.

The double-centerboard schooners never became common, but it was a style well known to Lake Michigan shipbuilders. When the lumber industry moved on to the Pacific coast, it brought the double-centerboard schooner. As on the Great Lakes, relatively few were built despite superior performance.

Why were so few double-centerboard vessels built? The added expense and complexity of construction may have been a factor. The reduction in capacity caused by the second centerboard also may have been a consideration. However, the principal issue may have been the level of seamanship required to safely navigate a double-centerboard schooner. In the cases of the *Lumberman* and the *Emeline*, at least, when the vessels were operated by competent captains with experience sailing the double-centerboard configuration, the vessels proved extremely fast and seaworthy. Fine-lined racehorses, they were prone to disaster if poorly handled. The *Lumberman* capsized under the command of a captain inexperienced with the rig. The *Emeline* suffered multiple capsizing events later in its career. In an industrial age when the sailor's arts were being eroded by the hard hand of industry, there was little room for these graceful but finicky racers.

The maritime history of the Lake Michigan lumber trade has been largely written from the perspective of its decline. In an influential doctoral dissertation completed at the University of Chicago and published by the US Congress in 1898, George Gerard Tunell described the still extensive Great Lakes lumber fleet as "very largely made up of vessels that are worthless for other purposes . . . most of the vessels of the old schooner fleet have been transformed into barges and are now engaged in the lumber trade."[55] Whether a small schooner picking up lumber from a mill located on shallow water or a larger schooner or steamer cut down into a tow barge, the lumber trade was the final stop in the careers of many commercial vessels on the Great Lakes. Broadly interpreted, the underwater archaeological record supports this observation. In Wisconsin's Lake Michigan waters, shipwrecks carrying cargo related to the lumber industry far outnumber those engaged in transporting any other type of cargo. And this was especially pronounced in the 1880s and 1890s, when well over half of all ships wrecking along Wisconsin's Michigan coast, including the *Lumberman* and the *Emeline*, were serving the trade.

The old trope of declining days of sail fails to do justice to the history of frontier ships and forests. The lumber schooners that continued to carry vast quantities of wood into the early decades of the twentieth century were not always old, nor were all decaying hulks repurposed for low-status trade. In his magisterial history of Chicago, *Nature's Metropolis*, William Cronon lays out the ecological and geographic determinants and organization of the lumber market at Chicago. Key to the rise was Lake Michigan's north–south orientation that "cut across—and so connected by water—radically different ecosystems," one with an extraordinary abundance of trees and the other virtually without forests.[56]

Mill owners, lumber dealers, and pioneer shipbuilders alike understood the significance of this natural arrangement. They sought ships that fit both the ecological and economic niche of the lumber trade to Chicago. In the 1850s and 1860s, the most common vessels designed and built for the lumber trade were the schooners and scow schooners. The *Lumberman* and similar double-centerboard schooners that followed represented the evolutionary apex of the lumber schooner on Lake Michigan. Later innovations associated with sailing schooners, such as the well-known Grand Haven rig introduced in the early 1870s, were adaptations of existing schooners to fit changing economic and technological circumstances, not fundamentally new designs, as had been the case with the schooners *Challenge*, *Magic*, and *Lumberman*.

The product of an old technology but part of a highly efficient system of production and distribution, the *Lumberman* left a considerable ecological footprint. In *Nature's Metropolis*, Cronon argues that the unexploited natural abundance of the frontier "offered to human labor rewards incommensurate with the effort expended in achieving them." In the lumber industry, these rewards could only be achieved with high-volume production and mass consumption. Reverend William Ferry, along with his children and grandchildren, generated their wealth through the system of production, transportation, and market distribution that transformed ecological riches accumulated over centuries into personal wealth. In the years of largest production, they did so without breaking a sweat. If Lake Michigan was both a corridor between two ecosystems and a "funnel" directing the natural wealth to Chicago, schooners such as the *Lumberman* were primary tools used to carry this wealth. And as the career of the *Lumberman* demonstrates, they could be powerful tools indeed. In twenty years under the ownership of the Ferry family, the *Lumberman* carried a conservative estimate of 100 million board feet of lumber. Every stick of wood moved from mill to market on the schooner and sold poured money into the pockets of the Ferrys and their associates. On an original investment of $10,000, the Ferrys secured a tool, the schooner *Lumberman*, that delivered to market for the family lumber and forest products with a conservative wholesale value of $1.5 million. As the Ferrys were lumber dealers as well as producers, the value carried was ultimately much greater.[57]

Chapter Seven

THE SCHOONER *KATE KELLY* AND ITS CAPTAIN

The story of the *Kate Kelly* actually consists of two overlapping tales. The first is a descriptive history of a representative Great Lakes canal schooner built in 1867 and lost in 1895; the second is the brief story of its final captain, Hartley J. Hatch—who was anything but typical for a Great Lakes mariner. The two stories, of the schooner and of the captain, overlap in the tragic foundering of the *Kate Kelly* during a violent blow on Lake Michigan on May 13, 1895.

Built in 1867, the two-masted 254-gross-ton schooner had a long and at times difficult career, principally sailing between Lake Ontario and Lake Michigan carrying coal west and returning with grain. The vessel was twenty-eight years old at the time of its loss, carrying a cargo of railroad ties. Worth an estimated $2,500, the schooner was not particularly valuable, and on the face of things, the *Kate Kelly* seems like more old schooner being worked to the bitter end. The story that unfolds in this chapter, however, is more ambiguous. In its final years the schooner *Kate Kelly* became something far more than a typical canal schooner, and if there ever was a Great Lakes mariner to bet on coming home from the sea, it was Captain Hartley J. Hatch.[1]

THE BUILDER

Born in Quebec in 1832 to French immigrants, John Martel moved to the United States in 1849 and eventually settled in Buffalo, New York, at that time the most important shipbuilding center on the Great Lakes.

He learned his trade patiently and well. By 1860, Martel, a mature ship carpenter, was married and had amassed personal property worth $300. By 1867, at age thirty-four, Martel had worked his way up through the ranks of competing artisans and began his career as an independent shipbuilder. The schooner *Kate Kelly* may have been his first vessel and was certainly one of his earliest. By 1870, by which point he had two young children, Martel had amassed real estate valued at $1,800 and personal property worth $400.[2]

By the time John Martel built the *Kate Kelly* in 1867, the economic conditions for independent shipbuilders in western New York and northern Ohio remained good but were on the decline. Two generations of shipbuilding and lumbering had largely consumed the local supply of shipbuilding timber. To continue converting trees into money (the central job of the frontier shipbuilder), many shipwrights moved west to Michigan and Wisconsin. In the early 1870s, John Martel joined this maritime migration and moved his family and business to Saugatuck, Michigan, a picturesque small town on the eastern shore of Lake Michigan.

Captain Hartley J. Hatch (1845–1895), inventor, master mariner, and the final owner and captain of the schooner *Kate Kelly*. COURTESY OF THE WHS MARITIME PRESERVATION AND ARCHAEOLOGY PROGRAM

In 1873 and 1874, Martel launched at least five schooners at Saugatuck, including the *Marinette, Menekomee, F. B. Stockbridge, George M. Case,* and *L. B. Coats.* The economic boom times did not last. A financial panic in the fall of 1873 ushered in a national depression that idled shipyards across the Great Lakes for several years. Martel, however, seems to have been more fortunate than many of his shipbuilding colleagues. He built two tugs in 1875 and another in 1877. Martel apparently understood that the days of building schooners were drawing to a close on the Great Lakes. When the region's maritime economy recovered at the end of the 1870s, he resumed building but focused on steamboats and became well known for his many tugs. Between 1880 and 1889, Martel launched forty-two vessels: five moderate-sized package/passengers boats, one schooner, and thirty-six tugs.[3]

While little is known of Martel beyond the ships he constructed, by all accounts, Martel was a stable and successful businessman of moderate means. After 1890, Martel's production slowed, but he continued to turn out the occasional tug and even a steam yacht. By 1900, census records suggest that the now widowed Martel had retired from active shipbuilding.[4]

The canal schooner *George M. Case* under construction in the Saugatuck, Michigan, shipyard of John Martel in 1874. ALPENA COUNTY GEORGE N. FLETCHER PUBLIC LIBRAY GREAT LAKES MARITIME COLLECTION

THE SHIP

The *Kate Kelly* was one of more than a thousand schooners designed to transit the second Welland Canal between Lakes Ontario and Erie between about 1845 and 1874. *Canallers*, as they were sometimes called, had to draw less than 9 feet of water when loaded and fit inside canal locks that were 150 feet long and 26.5 feet wide.[5] The design of canal schooners evolved over time, with shipbuilders developing adaptations to hull shape, rigging and davits to maximize the size of the hull capable of transiting the locks. These canallers took on an increasingly boxlike appearance as they evolved over the decades.[6]

The *Kate Kelly* was of moderate size for a canaller, measuring 126.3 feet in length, 25.8 feet in breadth, and 10.4 feet in depth. With a jibboom 45 feet long, the *Kelly*'s sparred length exceeded the canal lock dimensions by more than 20 feet, which meant that the jibboom had to be removed, or it might have been rigged to fold up and out of the way for the canal transit—an adaption found on some later canal schooners. As launched, the schooner reportedly spread two thousand yards of canvas sail on two masts. The mainmast was 22 inches in diameter and 84 feet tall, and the foremast was 24 inches in diameter and 82 feet tall. The main and fore booms were 60 and

50 feet long, respectively. Many of the canal schooners that came after the *Kate Kelly* were equipped with three masts and smaller booms and sails—a configuration that made sail handling much easier.

The living quarters of the new schooner were described as "finely fitted and furnished," with chestnut wood ceiling paneling. The aft cabin offered a compact 376 square feet of living space divided into seven compartments. The largest space, comprising 120 square feet, was the dining saloon, followed by the captain's cabin (80 square feet), the kitchen or galley (48 square feet), pantry (32 square feet), and four staterooms (24 square feet each) for the mates, cooks, and any guests who might make the voyage. Additional quarters for the common sailors were located in the forecastle in the bow of the ship. This was a typical arrangement for large schooners.[7]

THE OPERATIONAL CAREER OF THE *KATE KELLY*

During a career that spanned twenty-eight years, the *Kate Kelly* had several owners, most of them associated with trade on Lake Ontario. The schooner's career can be divided into four distinct periods: its first year sailing out of Buffalo (1867–1868), the Captain Robert Hayes era (1868–1877), the McFarlane/Goble partnership (1877–1893), and its final years based in Chicago with Captain Hartley Hatch (1893–1895).

Lake vessels represented a quick if volatile investment. Entire vessels and vessel shares frequently changed hands. John Martel built *Kate Kelly* for Louis Ryerse, a Buffalo ship owner. Ryerse kept the vessel for only a few months before selling it to Rumsen R. Brown and James M. Smith of Buffalo on September 30, 1867. In early April 1868, *Kate Kelly* again gained new owners when James Keller, Edward W. Parmalee, and Captain Robert Hayes purchased the schooner for $19,000. Hayes, who had a one-eighth share in the vessel, took command—a post he retained for nine years. The new owners changed the vessel's homeport from Buffalo to Oswego—the most important US port on Lake Ontario—and the *Kate Kelly* sailed out of Oswego for the next twenty-five years.[8]

Hayes and his associates employed the *Kate Kelly* to carry grain from western Lake Michigan ports, particularly Chicago, to Lake Ontario ports—most commonly Oswego and Kingston, Ontario. The vessel could hold approximately nineteen thousand bushels of wheat and more than five hundred tons of coal. While westbound from Lake Ontario, the schooner frequently carried coal and occasionally picked up odd cargoes such as railroad iron. The schooner sometimes made the westbound trip light and depended upon a good grain charter, which was common in the late 1860s, to make a profit. For example, in late October 1868, the ship earned nearly $3,000 for carrying one load of wheat from Milwaukee to Oswego at sixteen cents per bushel. At the beginning of the 1869 season, the schooner carried a cargo of corn from Milwaukee to Oswego at ten cents per bushel—generating almost $2,000 in freight revenue.[9]

Captain Hayes used the *Kate Kelly* hard. Newspapers, wreck lists, and admiralty court records report eight accidents and allude to others that involved the *Kate Kelly* between 1869 and 1877. The schooner sustained major damage in at least two of these incidents. In 1869, the vessel grounded near Cheboygan, Michigan, reportedly losing an entire cargo of wheat.[10] The 1871 *Classification of Lake Vessels and Barges*, an insurance register, noted that the *Kate Kelly* had undergone large repairs in 1870 and gave it just an A2 rating, an indication that the vessel was not in top condition for its age.[11] In April 1871, the vessel collided with the brig *Rosius* near the Eighteenth Street Bridge at Chicago.[12] In April 1874, it hit a dock at Oswego, and two months later, the battered vessel sprung a leak while transiting the Welland Canal. In September 1874, the schooner sustained damage colliding with a bridge at Toledo, Ohio, and rounded out the year by grounding at Ford's Shoal on Lake Ontario in November.[13] Collectively, the number and severity of the accidents call into question Captain Hayes's competence.

The year 1875 began better for Captain Hayes and the *Kate Kelly*. This all changed, however, in late September, when the vessel fetched up on shore about a hundred feet outside of the East Pier at Oswego. The vessel had been running light on a trip from Kingston, Ontario, and a stiff northwest wind had built up a large sea. As the *Kate Kelly* attempted to enter the port, however, the wind shifted to the southwest and nearly died out. Whether Hayes misjudged the wind direction or was surprised by its sudden moderation is unknown—but the vessel lost headway. Heavy seas carried the helpless schooner to the beach and deposited it broadside onto a hard bottom. The crew abandoned the ship, leaving it to pound on the beach throughout the night. By morning, a local newspaper reported that the *Kate Kelly* had broken its back, sprung its decks, and was holed.[14]

Initially, saving the schooner seemed unlikely, and had the vessel grounded this severely on a more distant shore, it would have proven a total loss. The *Kate Kelly*, however, was insured for $12,000, and Oswego possessed excellent ship salvage and repair capabilities. These factors encouraged the recovery of the heavily damaged craft. Insurance inspector Captain Berriman hired the well-known shipbuilder George Goble to assist

This bird's-eye illustration shows Oswego, New York, as seen from Lake Ontario in 1855. Connected to the main Erie Canal by its own canal in 1829, Oswego became the most important US port on Lake Ontario. Ships carrying grain from the western Great Lakes had to transit the Welland Canal to reach Oswego, which became a hub for the building and managing of canal schooners, including the *Kate Kelly*. LIBRARY OF CONGRESS PRINTS AND PHOTOGRAPHS DIVISION, 98508631

in removing the vessel from the beach. Goble and his partner, James D. McFarlane, operated one of the best-equipped shipyards on Lake Ontario. Using two powerful tugs, screw-jacks, and other implements, the salvers released the *Kate Kelly* from the beach. On September 30, 1875, the Goble-built tug *Alanson Sumner* managed to pull the schooner into the harbor, where it immediately sank. Three days, three steam pumps, and two canal boats later, the *Kate Kelly* floated again—this time just long enough to move it to the entrance to the Ontario Dry Dock, home of the Goble and McFarlane yard, where it sank a second time.[15]

As earlier reports had indicated, the *Kate Kelly* had sustained substantial damage. The keelson was broken aft of the centerboard trunk, several floor timbers were broken, and the starboard side had so many holes as to present "the appearance of a sieve." At first, the insurance appraisers could not agree on an estimated cost for repairs. An early observer predicted more than $6,000, while others suggested that the total bill, including salvage costs, would exceed the $12,000 in vessel insurance. Ultimately, Goble and

McFarlane installed fresh floor timbers and a new keel and replanked the bottom of the starboard side and some of the port side. They installed some new ceiling planking, pocket pieces on the centerboard trunk, and a new deck. The yard also lengthened the *Kate Kelly*'s bowsprit and added an additional jib. These additions to its headgear may have been intended to improve the schooner's handling. In any case, the vessel had only one more known collision after the yard altered its rig. It is not known who paid for the *Kate Kelly*'s repair, but Captain Hayes and his partners retained their ownership.[16]

The *Kate Kelly*'s many accidents raise questions about its management, which was, to put it mildly, fluid during the Robert Hayes era. Charles Parker, who owned the largest share of the vessel during the mid-1870s, according to the enrollments, died during the summer of 1874. The family's small shipbroker business fell into the hands of Parker's son, a man with a history of shady financial dealings. When the elder Parker died, the owners should have applied for a new enrollment for the *Kate Kelly*. They did not, an omission that suggests sloppy, if not dishonest, business practices.

Bad practices eventually caught up with the *Kate Kelly*'s owners. On October 22, 1877, William H. Wolf and Thomas Davidson, the operators of a large Milwaukee shipyard, filed a libel suit against the schooner for non-payment of debts. Wolf and Davidson contended that they were owed $875 plus interest for repairs and supplies they provided to the vessel after it had grounded yet again in October 1876. At the time of its arrest, the *Kate Kelly* was at Cleveland recovering from yet another mishap. On October 13, the schooner, carrying 380 tons of iron ore from Ogdensburg to Cleveland, ran aground on a sandbar. To float the vessel free, the crew threw 50 tons of ore into lake, worth well over $200.

The procedure for arresting and selling a ship at a federal public auction is popularly called a marshal's sale, a reference to the US Marshals who had responsibility of arresting the vessel in accordance with federal court orders. Captain Hayes, court records reveal, had left a stack of long-unpaid bills in Milwaukee and Chicago. When notice of the *Kate Kelly* arrest appeared in the newspapers, six other creditors presented bills. The cumulative debt against the schooner, including court costs, totaled $5,580.10, a figure considerably more than its current value. The largest claim against the vessel, $3,735.35, came from Goble and McFarlane, the shipbuilders who had rebuilt the *Kate Kelly* two years earlier. The Wolf and Davidson claim, including interest, came to $927.09. Other bills presented came from ship chandleries and a butcher for groceries and supplies, for trimming (loading) a cargo of grain, and from the owners of the recently jettisoned iron ore.[17]

With pending bills far higher than the *Kate Kelly*'s value, its owners made no effort to pay off the creditors. Court records suggest that problems plagued the *Kate Kelly* partnership. The younger Parker, half owner, appears to have been insolvent by this time, and Henry W. Green, owner of a quarter share, indicated that "he might secure his own interest, but would do nothing to benefit Parker." For his part, Captain Hayes, who owned the final one-quarter of the vessel, admitted that the claims against the vessel were just and accurate but indicated that he was "unwilling and unable" to pay the creditors. The vessel went on the auction block on November 6, 1877, and was sold to George Goble and James McFarlane for $3,000.[18]

The question arises as to why these stable, first-rate businessmen with excellent credit and good judgment purchased the benighted *Kate Kelly* during a tough period in the Great Lakes economy. The answer lies in the vagaries of Admiralty Law. By initiating the arrest of the *Kate Kelly*, Wolf and Davidson became the first parties in line for the disbursement of auction proceeds. They received their full claim, both principal and interest. After deductions for $351.65 in court costs, the other six creditors had to split the remaining $1,671.16, with each receiving just 39.5 percent of their original claims. Under these rules, Goble and McFarlane recovered $1,475.46 of the $3,000 purchase price. By putting up an additional $1,524.54, Goble and McFarlane prevented the irrecoverable loss of more than $2,200 on their $3,735.35 claim against the *Kate Kelly* and gained possession of a fully serviceable schooner.[19]

The admiralty court files preserve many of the original documents submitted by the plaintiffs, including grocery bills—particularly significant to historians of Great Lakes seafaring life. Sailor's lore praised the plentiful food and fresh water aboard Great Lakes ships. The lore seems to be borne out by fact on the *Kate Kelly* in the early winter of 1876. Before setting sail from Milwaukee with 15,500 bushels of wheat, the vessel took on $83.53 worth of food. The crew appears to have subsisted on large quantities of salted and fresh meat. The ship left port with 55 pounds of salt pork, 153 pounds of corned beef, and 215 pounds of various cuts of fresh meats. This supply would have provided a typical crew of eight persons with well over two pounds of meat for each hand, every day, for a three-week period—a time span longer than the typical voyage between Milwaukee and Oswego or Kingston. Adding to the protein component were ten pounds of salted mackerel. The cook also kept busy baking, and the ship took on thirty pounds of sugar, twenty-five pounds of butter, ten pounds of lard, three-fourths a barrel of flour, six dozen eggs, and one and a half bushels of apples.

Other starches came from two bushels of potatoes and a half bushel each of turnips and beets. The only green vegetables in evidence were several cabbages. Small quantities of rice, beans, pickles, and crackers provided additional diversity to the diet. Spices and condiments were simple and few, just salt and mustard. Great Lakes coffee, a favorite sailor's beverage, must have been extremely watery as the ship took on only six pounds for the voyage. The ship purchased no other beverages, but fresh water was always plentiful.[20]

The *Kate Kelly*'s new owners, George Goble and John McFarlane, had little in common with Robert Hayes and his associates. The Ireland-born Goble was one of Lake Ontario's most respected shipbuilders and had been a fixture on the Oswego waterfront since the late 1830s, while McFarlane's father-in-law was F. G. Carrington, Oswego's most prosperous business-man.[21] The two successfully managed the vessel for more than fifteen years. During the 1880s and early 1890s, the *British Whig*, a Kingston newspaper, intermittently recorded the schooner's travels, cargoes, and occasional problems. During the Goble and McFarlane era, the *Kate Kelly* made frequent trips carrying grain—corn or wheat—between Kingston, Ontario, and Chicago. The vessel seems to have generally stayed busy and, when rates were high, must have shown a tidy profit. For example, in August 1883, the vessel grossed $1,000 for carrying sixteen thousand bushels of wheat from Chicago to Kingston. In the summer of 1888, the schooner was grossing about $600 per trip on much shorter voyages carrying coal between Sandusky and Kingston.

Oswego was thriving in the mid-1880s with growing wealth being chan-neled into new leisure activities—among them yacht racing. In 1886, a group chartered the *Kate Kelly* as a platform to watch and celebrate the annual Oswego Regatta. The ship hosted "a large party of ladies and gentlemen." With the decks cleared for action, the group watched the races, ate a fine meal, and danced waltzes on the pine decks to the accompaniment of an excellent band. "In short," the *Oswego Daily Times* reported, "everything combined to make the day a memorable one."[22]

Even under solid management, sailing ships on the Great Lakes faced constant dangers. As ships aged, the potential for problems increased. In September 1881, the *Kate Kelly* lost its fore boom during a storm. On November 13, 1882, the vessel ran aground during a blow in Lake Ontario. A rescue party of thirty men and a powerful tug lightered five thousand bushels of wheat from the schooner and pulled it from the shore. The vessel underwent an extensive rebuilding at the Goble yard in 1883. In 1886, part of its foremast and its maintop mast tore away during a June squall. On November 11, 1891, a vessel struck the *Kate Kelly*, seriously damaging the stern.[23]

The sailing life was dangerous for people as well as ships, and in 1890, James O'Hara, a forty-year-old sailor from Rochester, fell from the main boom on the *Kate Kelly* while furling a sail in Detroit and died. In May 1892, Captain Robert Hayes, the schooner's former part owner and master, fell overboard from the schooner barge *Ogarita* and drowned in Lake Erie.[24] By the early 1890s, large wooden and steel steam-powered vessels, many exceeding 300 feet in length, had taken over the major Great Lakes bulk cargo trades, and many schooners had their rigging cut down for use as tow barges. By 1891, the *Kate Kelly* retained its full sailing rig but was typically collecting $300 or less in freight fees per trip carrying grain at one and one-half cents per bushel, barely enough to cover expenses.[25] In March 1893, Goble and his partner sold the *Kate Kelly* to Captain Hartley Hatch of Chicago for an undisclosed sum.

INNOVATION IN THE CLOSING DAYS OF SAIL

In an age when shipwreck and death on the water still happened on a routine basis, the loss of an entire crew and captain, while worthy of remark and sympathy, rarely sparked the kind of reaction exhibited in the newspapers by Chicago's maritime community following the loss of Captain Hatch and the *Kate Kelly*.

The newspaper accounts of the wreck of the *Kate Kelly* described Captain Hartley J. Hatch as "one of the oldest navigators on the Great Lakes." They referred to his commanding multiple voyages to salt water on Great Lakes vessels, including seemingly wild tales of reaching the Mediterranean Sea and the Indian Ocean. Six days after the loss of the *Kate Kelly*, the front page of the May 19, 1895, issue of the *Chicago Inter Ocean* newspaper opened the top of a long story with these heading lines:

> HATCH HAD A HISTORY. Captain of the Ill-fated Kate Kelly Widely Known. HE SAILED IN EVERY SEA. Was a Mariner of Acknowledged Skill and Experience.[26]

Accompanying the story were three large woodcut illustrations depicting the search for the *Kate Kelly*, another unusual response for the loss of a common schooner. Hartley Hatch, however, was no common seaman. Nor, as it turns out, was the *Kate Kelly* a typical canal schooner when it sank.

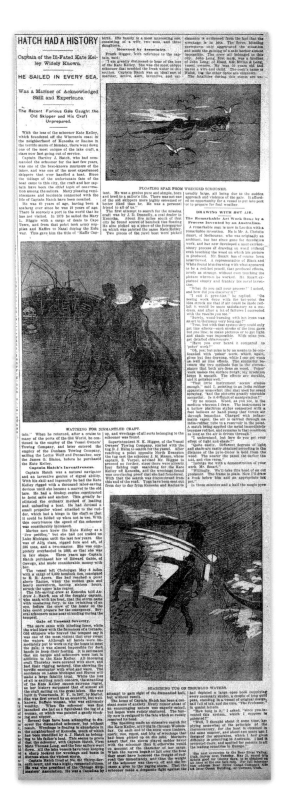

Hartley Hatch was born in Woodstock, Ontario, in 1845 and began sailing the lakes in the early 1860s. Settling in Chicago, Hatch married and became the father of five children, including three daughters who were still living at home when the *Kate Kelly* sank.

When Hatch began commanding vessels is unknown, but in September 1877, he was the captain of the large Great Lakes schooner *Mary L. Higgie*, then on the final leg of the first of three trans-Atlantic trading voyages. The *Mary L. Higgie* was a canaller built in Manitowoc, Wisconsin, in 1872 by Jasper Hanson and Hans Scove, a pair of Danish-born shipbuilders who immigrated to the Great Lakes during the 1850s. The 310-gross-ton topsail schooner was 139 feet long, 26.3 feet in breadth, and 11.2 feet in depth.

Hatch's second trans-Atlantic voyage commenced in October 1877, with the *Higgie* sailing from Montreal to Cape Town, South Africa, with a cargo of timber. A surprised Illinois traveler found the schooner anchored at Table Bay and reported the event in a letter to his hometown newspaper in Alton, Illinois:

> A short time ago I was very much surprised to see at anchor in Table Bay, a lake Michigan schooner, the *Mary L. Higgie*, owned in Chicago and commanded by Capt. Hatch. The schooner went round the lakes from Chicago to Quebec, loaded there and then came to Cape Town. She has now gone to East London with a freight for the government, and if she gets out of that open roadstead alright, her owners are not likely to regret having sent her to the

The *Inter Ocean*'s extensive illustrated coverage of the loss of the *Kate Kelly* and her crew was unusual, revealing the high esteem for Captain Hatch in the Lake Michigan maritime community. CHICAGO *INTER OCEAN* NEWSPAPER, MAY 19, 1895

Cape, as she got a splendid freight from here. She is the first vessel I have ever seen here without copper; in lieu of that her bottom is painted with some composition or other, that is supposed to substitute for the metal.[27]

While in South Africa, Captain Hatch got caught up in the British colonial wars when British officials chartered his ship to carry naval stores from Cape Town to East London, a port on South Africa's Indian Ocean coast about eight hundred sea miles to the east. At East London, the *Higgie* was contracted to carry a contingent of native African fighters who were supporting the British to Algoa Bay. The conflict ignited when, after decades of British occupation, the Zulu raised an army and rebelled. The chartering of Hatch's Great Lakes schooner to carry pro-British African warriors was one of the opening acts that led to the outbreak of the Anglo-Zulu War in 1879.

At Algoa Bay, Hatch delivered the warriors, picked up a load of South African wool, and set sail for Montreal, arriving after a rough voyage in mid-October 1878. Hatch made one more Atlantic voyage before bringing the *Higgie* back to the Great Lakes, delivering a cargo of Spanish salt and a deck load of Michigan lumber to Chicago in 1879.[28] Although the *Higgie* is an extreme example, other Wisconsin-connected sailing vessels made long

The 1878 South African voyages of the Great Lakes schooner *Mary L. Higgie*. DAVID RUMSEY MAP COLLECTION

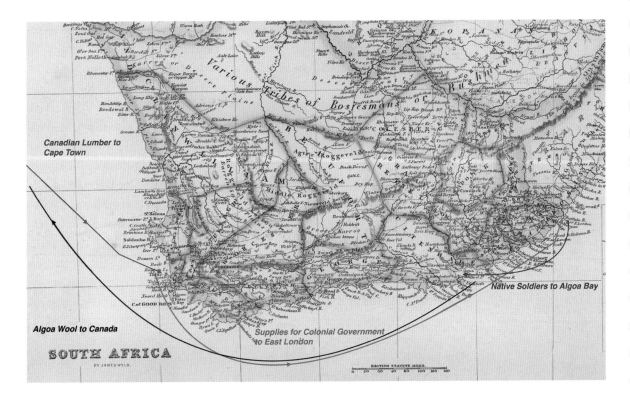

trans-Atlantic voyages during this period. In 1878, the canal schooner *Floretta* sailed with a load of timber from Grand Haven, Michigan, to Glasgow, Scotland. The vessel returned to the Great Lakes in 1879 after stops at British Guiana, Barbados, and Montreal. The *Floretta* sank off Manitowoc with a cargo of iron ore in 1887. The wreck was documented by Wisconsin Historical Society archaeologists in 2013.[29]

Hatch and his employers, the Higgie family, made it through the dark economic years of the late 1870s through audacious international trading ventures and Hatch's excellent seamanship. With the increasing prosperity on the Great Lakes evident in 1879, the *Mary L. Higgie* and Hatch quit the ocean, returning to full-time freshwater sailing.[30] Hartley Hatch left the Higgie family's employment and spent some years as the captain of the *Viking*, a famous sailing yacht brought to the Great Lakes in 1882 by Chicago lumber baron Colonel John Mason Loomis. Following that, he moved to the Vessel Owners Towing Company controlled by the Higgie family and then took a position with the Chicago tugboat king J. S. Dunham, sailing the *Pensaukee*, one of Dunham's large schooners.[31]

In 1891, Captain Alexander McDougall, inventor of the "whaleback" steel freighter and owner of the American Steel Barge Company at Superior, Wisconsin, was looking to expand his enterprises to the Atlantic and Pacific Oceans and had established a shipyard at the Puget Sound city of Everett, Washington. To create public interest in his unique cigar-shaped ships, on June 11, 1891, McDougall dispatched the new whaleback steamer *Charles W. Wetmore* on a highly publicized international voyage, first to Liverpool and then to New York, Philadelphia, and on to the US Pacific coast.[32] Hartley Hatch sailed on the *Wetmore* voyage, but his position is unknown. He was a highly respected mariner, and his knowledge of the St. Lawrence River and English ports and experience with foreign shores and the protocols of international shipping would have proven invaluable. The *Wetmore* made the passage from Montreal to Liverpool in eleven days carrying ninety-five thousand bushels of wheat and recrossed the Atlantic to New York City in fourteen days. On the final leg of the journey from Philadelphia to Everett, Washington, the *Wetmore* encountered a storm off the Oregon coast that was described as the worst in fifteen years. The ship lost its rudder, and the crew survived the storm by jury-rigging masts and deploying sixty fathoms of anchor chain as a sea anchor before being found and towed in to Astoria, Oregon, for repairs. The ship and crew had a narrow escape. After the *Wetmore* reached Puget Sound, Hatch returned to Chicago to work for the Vessel Owners Towing Company.[33]

Views of the whaleback *Charles Wetmore* on the transatlantic voyage to Liverpool in 1891. *HARPER'S WEEKLY*, SEPTEMBER 5, 1891

By 1893, Captain Hartley J. Hatch of Chicago had commanded sail and steam vessels on the Great Lakes for twenty years or more, including some of the most prosperous years of the wooden age. While other entrepreneurial mariners had risked their life and capital to accumulate wealth and achieve independence, Hatch had always remained in the employment of others—taking the physical risks, spending years away from family, but not reaping the large financial rewards. He was a respected, capable, and resourceful mariner but not a wealthy man. Perhaps it was a deep desire on the part of an ambitious and talented forty-seven-year-old mariner to finally achieve independence that led him to purchase the schooner *Kate Kelly* and bring it to Chicago in March 1893.[34]

With the national economy beginning a freefall into the Panic of 1893, this was a tough time to own any ship. With his connections, however, Hatch may have been well positioned to get good charters for the old but inexpensive to operate *Kate Kelly*. In early May, he earned between $350 and $400 carrying coal from Sandusky, Ohio, to Goderich, Ontario. During summer and early fall, the schooner made multiple trips carrying oats from Chicago, sometimes to Lake Huron but also to Kingston, a Canadian port on Lake Ontario.

In October 1893, Captain Hatch arranged to carry a heavy cargo of valuable grindstones from Grindstone City, Michigan, to Milwaukee. A vicious gale came up during the loading and Hatch deliberately sank the *Kate Kelly* at the dock to save it from pounding to pieces in surf. Coming in to Lake Michigan, the *Kate Kelly* ran aground on Spider Island near the tip of the Door Peninsula. The *Door County Advocate* marine reporter complained that the vessel's cargo, worth $10,000, was "far too valuable to risk out in an old leaky vessel at this season of the year." The *Kate Kelly* was indeed old, but it had been kept in good repair by Hatch and the previous owner, shipbuilder George Goble. Captain Hatch refloated the schooner and had it towed to Milwaukee, where new bottom planking, garboard strake (the critical plank next to the keel), and forefoot were installed at a cost of approximately $1,100—nearly half the vessel's estimated market value.[35]

Hatch scratched away with the old schooner throughout the 1894 season. On May 1, he received a charter to carry a cargo of oats to Port Huron, Michigan, for 1¼ cents per bushel, a trip that probably grossed $250 to $300 in freights.[36] Chicago oats continued as the schooner's principal eastward cargo during the summer of 1894. The *Kate Kelly* picked up a load of coal at Sandusky, Ohio, in September and probably made other westbound coal trips as well during the season.

FATE OF THE *KATE KELLY*

On Thursday, April 26, 1895, the *Kate Kelly* cleared the port of Chicago bound for Cheboygan, Michigan. Captain Hatch had received a charter to pick up a load of hemlock rail ties. Reports of the cargo's size range from forty-five hundred to six thousand ties. This translates to somewhere between 300 and 450 tons of weight and roughly twenty thousand cubic feet of cargo (between seven and nine and half standard 400-foot-long shipping container boxes). That is a lot of volume to pack inside the hold of a 125-foot-long schooner, and a large number of ties were carried as a deck load. When the *Kate Kelly* arrived at Cheboygan or how many days it took to load the ties is unknown, but on May 9, the schooner cleared Cheboygan bound for Chicago. The weather was in the balmy sixties, winds light, barometer rising—it looked as though they would have a slow but pleasant voyage home. The following day, Chicago and Milwaukee enjoyed unseasonably warm seventy-plus-degree weather, but on the northern edge of the lakes, a cold front was bringing temperatures down. Marquette, Michigan, reported a high of seventy-six on Thursday morning but only fifty-two degrees on Friday morning. The weather in northern Lake Michigan and the Straights of Mackinaw was overcast and rainy with variable winds. By Saturday, the cold front intensified, bringing temperatures on the lake down into the thirties with a mixture of snow and heavy rain and winds on the northern part of the lake.[37]

Sunday, May 12, found the *Kate Kelly* sailing in clearing but cold weather as a strong high-pressure system began moving in from the southwest, pushing out the moisture but bringing heavy winds. The *Chicago Daily Tribune* reported the "chill north winds" as a "godsend" to the lumber-laden sailing vessels on Lake Michigan that had been beating against shifting winds from the south for several days. "The change to the north was a signal for each Captain to begin to get all the canvas on his vessel and make a race for the lumber docks in the Chicago River."[38] Farther to the north, the cold front brought temperatures at the Sault Locks down to the teens, with extensive killing frosts damaging the crops of fruit growers in Wisconsin and Michigan. On southern Lake Superior and northern Lake Michigan, large steamers and schooners alike were getting rough treatment. The 450-ton steam barge *Philetus Sawyer* lost a deck load of stone in heavy seas between Door County and St. Joseph, Michigan. The *Kate Kelly*, with or without its large deck load, would have been having a difficult time.

On Monday, May 13, the US Weather Bureau's 8 a.m. national map depicted a strong and unexpected low-pressure center over southern Lake

Michigan that created the conditions for an explosive spring storm. By Monday morning, the *Kate Kelly* was laboring in heavy seas about two miles off of Wind Point, near Racine, Wisconsin. On shore, farmers watched as the schooner suddenly capsized and four men climbed into the rigging. Moments later, the ship disappeared below the surface. The farmers notified the local lifesaving crew, who could do nothing to assist because of the terrible weather. The farmers had observed that the schooner was rather large and had two masts.[39]

The storm had been unexpected and severe. Willis Moore, the new chief of the US Weather Bureau, made what was for him a rare admission of surprise, confessing "that on Sunday we had no idea that within twenty-four hours such a storm would be upon us. . . . At 8 o'clock last night there was no indication of a storm center anywhere in the United States." In Chicago, the report of the two-masted schooner capsizing led to immediate concerns for the safety of the *Kate Kelly*. By Tuesday afternoon, the fears were confirmed when reports of vessel debris, including a name board and hundreds of rail ties, made it clear that the *Kate Kelly* had indeed gone down. Knowledgeable mariners speculated that the schooner's heavy deck load shifted, causing the ship to lose equilibrium and capsize.[40]

On Wednesday, the local US Life-Saving Service crew spied a mast protruding several feet out of the water. It proved to be the *Kate Kelly*. On June 9, 1895, diver John Harms made three exploratory dives on the wreck,

This mast step rested on the main keelson and held one of the *Kate Kelly*'s masts. TAMARA THOMSEN, WHI IMAGE ID 140121

resting in fifty-five feet of water. He determined that the wrecked schooner was pointed to the northeast. Walking along the bottom he observed extensive damage to the hull; part of the decks and rail were gone, but the jibboom and bowsprit were intact. The foremast was gone, one anchor had been cast overboard, and the mizzenmast was enveloped in a tangle of rigging. Harms found no bodies but brought up a large section of flag that had been placed at half-mast in the forward rigging—a last but futile attempt at a distress signal.[41]

Given the eyewitness accounts, the distress flag, and the reported condition of the wreck, it seems likely that the *Kate Kelly* battled the weather for some time. The end, however, came quickly, with the vessel either capsizing or plunging to the lake bottom after a large sea or set of seas swept the deck with enough force to tear off the cabin, yawl boat, monkey rails, and anchor. The *Kate Kelly*'s remaining mast posed a navigation hazard and was removed from the wreck in late July. Possibly as a consequence, a few days after the mast was removed, three bodies floated ashore, including that of Captain Hatch. This suggests that the men were trapped when the ship rapidly went down.[42]

By the time of its loss, Captain Hatch had made important technological adaptations to the *Kate Kelly* that reduced the required labor and operating costs of the old schooner. Reducing crew size was one of the myriad ways vessel operators cut costs to remain profitable. On the largest class of schooners and schooner barges, steam donkey engines became increasingly common in the late nineteenth century to assist with raising sails and pulling anchors. In 1895, however, they were almost unheard of on an old canal schooner. However, the *Kate Kelly* had a donkey engine and more. Unlike most later canal schooners, the *Kate Kelly* had two rather than three masts. Canallers with a two-mast rig employed larger sails that required more manpower to raise and handle, particularly in bad weather. Using the donkey engine and unspecified rigging innovations mentioned in newspaper accounts, Captain Hatch was able to run his schooner with one highly experienced mate, fifty-year-old John Long, three seamen, and a cook—two to three fewer hands than the normal complement for a schooner of its type.

Frank Higgie, Hatch's former employer and close friend, lamented the loss of the *Kate Kelly* and its captain:

I am greatly distressed to hear the loss of the *Kate Kelly*. It was the most unique schooner that scudded freshwater in this section. Captain Hatch was an ideal sort of mariner, active, alert, inventive, and valiant.

He was a genius pure and simple, born and bred to a sailor's life. There was not one of the old skippers better liked then he. He was a personal friend to all of us.[43]

The marine reporter for the *Chicago Inter Ocean* described some of Hatch's unique additions to his schooner: "He had the *Kate Kelly* rigged with a thousand labor-saving devices until she became a marvel to the old tars. He had a donkey engine constructed to hoist sail and anchor. This greatly facilitated the loading and unloading a boat. He had devised a small propeller wheel attached to the rudder, which had a hinge in the shaft that it could be folded up when not in use. With this contrivance the speed of the schooner was considerably increased."[44]

In 2002 and 2003, volunteer divers from the Great Lakes Research Foundation, Inc., and archaeologists from the Wisconsin Historical Society surveyed the wreck of the *Kate Kelly*. The survey included a standard non-disturbance mapping of the site and documentation of construction features and artifacts.[45] The schooner's hull had been seriously damaged during the wrecking event and further compromised by the later removal of the mizzenmast. The wreck is broken into several sections with the most intact being a 112-foot section of the bottom. Half of the ship's centerboard remains inside the partially intact centerboard trunk. The other half is broken cleanly away and lies on the bottom on the starboard side of the hull. This resulted from the later damage rather than from the wreck event. Port and starboard bow sections have broken away, with the vessel's windlass resting on top of one of them. Forty-six feet of anchor chain stretches out from the bow along the lake bottom, suggesting that Hatch had dropped an anchor prior to the ship's sinking.

The principal evidence remaining on the site of Hatch's unusual incorporation of steam is a 16-foot steel shaft connected to 10-inch and 40-inch diameter gears. Attached on each end of the shaft are winch heads, both 5 inches in diameter. Wide slots have been cut into each head, probably a modification made to improve the handling when the running rigging (rope) to the sails or handling cargo. Fragments of the boiler remain on the site, as does one of the schooner's cast iron double-acting bilge pumps. Off the stern of the vessel, a surviving lumber port cover suggests that Hatch or one of his predecessors modified the hull to carry wood, as lumber port covers are not features of standard grain and coal schooners.[46]

Archaeologists found no evidence of special propulsion machinery or other innovations, and the donkey steam engine was not on the site. These

This gear and shaft assembly channeled power from the *Kate Kelly*'s steam engine to labor-saving devices that assisted with sail and cargo handling. Captain Hatch also installed a one-of-a-kind steam-powered auxiliary propulsion system. TAMARA THOMSEN, WHI IMAGE ID 140119

The steam engine of the *Kate Kelly*, on display at Lockwood Pioneer Scuba Museum in Loves Park, Illinois. Very few old canal schooners added steam engines, and the *Kate Kelly* is the only one known to have employed one for auxilliary propulsion. JOHN JENSEN

had been removed by shipwreck hunter and diver Dan Johnson, who found the wrecks of both the *Lumberman* and the *Kate Kelly* in the early 1980s. This was some years before the passage of state and federal shipwreck protection laws. Many of the artifacts, including the donkey engine and parts of the propulsion system, are on display along with more typical shipwreck artifacts at the Lockwood Pioneer Scuba Museum in Loves Park, Illinois, owned by Johnson. Analyzed as a collection, the *Kate Kelly* artifacts may have more to reveal about the *Kate Kelly*'s final years afloat and of the inventive mind of Captain Hartley Hatch.

While Hatch had not been an entrepreneur as a young man, or at least not a particularly successful one, he had indeed been ambitious and inventive. On April 30, 1872, he was issued US Patent number 126,204, *Improvement in Canal-Boats*. The patent was for a self-propelled canal boat with a detachable pointed bow and a hinged removable paddle-box drive system. Hatch's design was intended to improve the performance of canal boats and reduce their wake when under way without loss of capacity or ability to transit standard locks. This seems to have been Hatch's only patent, but when he died, he had a well-established reputation for inventiveness among his peers, and was no mere tinkerer.[47]

Captain Hartley Hatch was, however, a victim of bad luck and bad timing. Schooner captains had fewer options when caught in sudden storms than did the commanders of steam vessels. The winds in the two days leading up to the foundering of the *Kate Kelly* offered Hatch little chance to reach Chicago. The shoreline below the Door Peninsula offers no natural harbors of refuge, and by the time he reached the southern half of Lake Michigan, high winds and heavy seas made entering any of the few harbors a perilous enterprise. Captain Hatch probably made every good decision possible given the situation, but he ran out options and lost his ship and his life.

His timing had also been poor in buying the schooner in the spring of 1893, and it was a testament to his ingenuity and status within the Chicago maritime community that he had made money with the *Kate Kelly* despite the depressed economy and falling freight rates. Had the accident

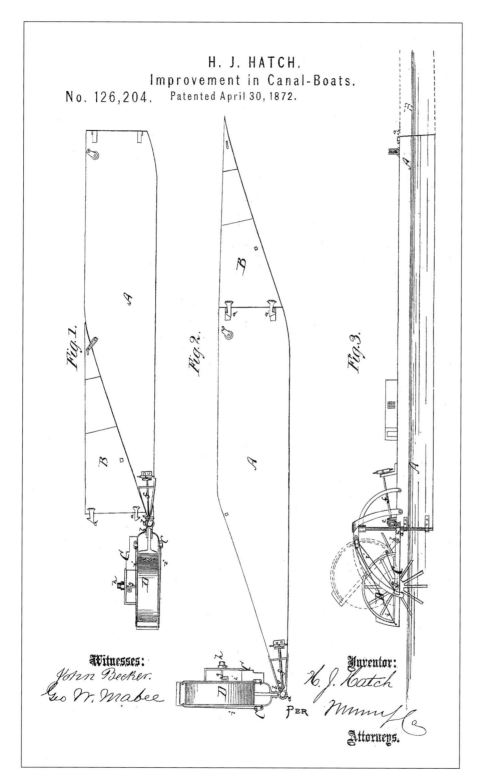

not occurred, Hatch may have enjoyed some very profitable years around the turn of the century when demand for ships to carry coal and iron grew insatiable and freights became very high. Other schooners, such as the famed Christmas tree ship *Rouse Simmons*, found niches that carried them profitably into the twentieth century. Hatch was a mariner of probably unmatched experience among his peers and was clearly an admired, even beloved member of his community. If anyone could have made a success of the waning days of wood and sail on the Great Lakes, it was Hatch.

In the final analysis, the *Kate Kelly* was not a typical old schooner driven to the bitter end. It was a bridge, or perhaps more accurately, a lifeline between a receding Atlantic maritime culture of canvas, wood, personal relationships, and independence that had characterized the Great Lakes during its frontier period into the mid-1800s and the modern corporate-dominated, capital-intensive industrial waterway it was fast becoming. Hartley Hatch had sailed across a third of the world's oceans, taken part in international trade as well as a colonial war, and witnessed the rapid transformation of his home region. Like Charles Reed, Allyne Litchfield, and later Captain James Davidson, Hatch was, or at least became through his maritime occupation, a cosmopolitan—a person possessing wider world perspectives beyond the ordinary in late nineteenth-century America. As a wreck site, the *Kate Kelly* represents the optimism, the creativity, and the tragedies forged by the fusion of Atlantic maritime culture and freshwater navigation during the wooden age on the Great Lakes.

Chapter Eight

THE EDUCATION OF CAPTAIN JAMES DAVIDSON

In the decades leading up to the Civil War, abundant Great Lakes forests supplied inexpensive fuel for steamboats and seemingly limitless quantities of reasonably priced, high-quality shipbuilding material. Aside from a few notable exceptions, wood remained the shipbuilders' material of choice throughout the United States, including on the Great Lakes. Also in these earlier years, many Great Lakes shipbuilders, ship captains, and common sailors began their careers at saltwater ports. The influence of Atlantic maritime culture on sail and steam vessels and on the organization of maritime labor and commercial practices on the Great Lakes during the early and mid-nineteenth century is obvious.

By the turn of the twentieth century, however, the Great Lakes had become a more distinctive maritime region, with unique technologies afloat and ashore and an increasing percentage of its working population working their entire careers on the freshwater seas. Although some wooden sailing ships and many wooden steamers continued in service, by 1900 huge steel bulk carriers dominated the lakes. These industrially produced cargo-moving machines met the needs of America's fast-growing industrial economy. The operating of ships by individual small-business owners or owner-captains was disappearing. To compete on the Great Lakes in the twentieth century required extensive capital and knowledge of business. Industrial capitalism and the corporation, not the maritime entrepreneur of old, held sway. Maritime culture, however, can be highly adaptive, and a few of the old-style mariner-businessmen negotiated the jarring transitions

A rare image of Captain James Davidson (1841–1929). COURTESY OF THE WHS MARITIME PRESERVATION AND ARCHAEOLOGY PROGRAM

that transformed the Great Lakes in the years between the end of the Civil War and the dawn of the new century. One the most successful of these men, and possibly the most colorful, was Captain James Davidson (1841–1929).

Wisconsin's collection of historic shipwrecks provides a vast material record of the transformation from a wooden age frontier to maritime region dominated by industrial capitalism. At least eight Davidson-built vessels were wrecked or abandoned in Wisconsin, and all or nearly all of the ships he built or owned passed through state waters. Davidson's story, and those stories embodied in the preserved wrecks of Davidson-built vessels, provide a touchstone and voice to the dynamic history of the Great Lakes from the maritime frontier of the palace steamer era of the 1840s and 1850s through its evolution into the most efficient industrial waterway in the world.

Beginning in the early 1870s, Captain James Davidson launched more than one hundred wooden vessels from his shipyard in West Bay City, Michigan, among them several of the largest wooden steamers and sailing ships to ply the Great Lakes. By continuing to build and operate large wooden ships into the early twentieth century, Davidson managed the transition from the wooden to the industrial age on the Great Lakes with more success than almost any mariner-entrepreneur of his generation. While so many mariners and shipbuilders from his time died, went broke, or ended up working for others, Davidson grew richer and more powerful, but above all remained independent to the very end. His success as a shipbuilder, ship owner, and industrial capitalist was no accident; it was the result of Davidson's innate drive and talent, personal characteristics that he carefully honed through an extraordinary process of practical education.

THE EDUCATION OF A SEAFARING MAN

James A. Davidson was born in Buffalo, New York, on August 12, 1841. His parents, Joseph and Elizabeth Davidson, had emigrated from Scotland and settled in the thriving frontier port of Buffalo sometime in the later 1820s. The elder Davidson participated in the construction of the Buffalo harbor's long breakwater, a protective structure unique on the early Great Lakes for its extensive stone armoring. The project was initially completed in 1828, but improvements, including the stonework, continued into the early 1830s. Joseph Davidson apparently subcontracted at least some of the stonework.

Divers investigate wreckage of the bulk steamer *James Davidson* at the Thunder Bay National Marine Sanctuary in Lake Huron. NOAA, THUNDER BAY NATIONAL MARINE SANCTUARY

The complexity of the project and its success as a structure raise the possibility that Joseph Davidson brought some knowledge of coastal engineering from Scotland to the Great Lakes. Whatever his skills and his contributions to building the Buffalo harbor, the elder Davidson seems to have left no traces in the historical record of the city. We can assume that Joseph Davidson was well known in the small social world of a frontier port and that his son had an early introduction to Lake Erie and its maritime culture.

James Davidson was born in the midst of a prolonged depression, but he grew into his teens during an exceptional period of opportunity, optimism, and economic expansion. On Atlantic and freshwater coasts of North America, booming trade sparked what historians remember as the golden age of American shipbuilding. Blessed with vast forest resources, shipbuilders in the northeastern Atlantic and Great Lakes regions of the United States launched thousands of ships during the 1840s and 1850s. Locally, Davidson undoubtedly witnessed the launching of palace steamers such as the *Niagara* and its larger cousins. We can imagine the young boy walking with his father along the stone breakwater, discussing ships and picking up some finer points of stonework and coastal engineering. We can only speculate what Davidson might have learned from his father. However, Buffalo offered undeniable opportunities for a highly gifted and ambitious boy to become enmeshed in the Atlantic maritime culture as well as the freshwater frontier.

Epidemic diseases periodically ravaged the United States during the 1840s and 1850s, with the frontier ports on lakes and rivers proving especially vulnerable. Both of Davidson's parents died in 1852, possibly from cholera. His biographical statements do not indicate whether he went to live in an orphanage, or if someone in his family took him in, but it is known that James went to work immediately after his parents died. According to Mansfield's *History of the Great Lakes*, published during Davidson's lifetime, "of necessity he became self-supporting" at age eleven and established a ferry across the river at Buffalo.[1] In a version told by another historian, the young orphan built a rowboat and carried sailors out to anchored ships.[2]

Whatever the specific circumstances, multiple accounts agree that Davidson's first job at age eleven was ferrying people from a landing at the foot of Main Street to ships in the Buffalo harbor. The work placed young Davidson in direct contact with sailors from all parts of the Atlantic. Orphaned and ambitious, he would have soaked up the seafaring culture, listening attentively to colorful sailor's yarns. While the entrepreneurial parts of the Davidson ferry story may seem a bit stretched, there is no doubt the boy demonstrated a lot pluck and became a familiar figure on the Buffalo waterfront. In addition to his maritime work, the boy found time for school. Whatever challenges he faced as an orphan, Davidson was lucky to have been born in Buffalo, a city that maintained a system of free public schools that by 1852 included both an evening school and a high school. At a time when most nonelite American men had at most a few years of elementary school, Davidson took every advantage of the opportunities for formal education the thriving city offered. Clearly a maritime model of the self-made man, Davidson's fusion of vocational, experiential, and formal education equipped him with professional tools and economic resilience matched by few, if any, of his lake-faring contemporaries.

Davidson's tenure as a ferry boy may have been brief, or at least intermittent; he reported first sailing the lakes in 1853, at the age of twelve. According to the Mansfield biographical sketch, Davidson shipped out as second mate on a schooner at age seventeen in 1858 and as a master during 1860 season. The rapid ascent to officer status is impressive but not that unusual in 1850s maritime America, when rapidly growing fleets of small coasting schooners needed competent officers willing to work for low wages. What is remarkable is that from an early age, Davidson prepared for a career more ambitious than commanding a schooner.

In the late 1850s, Davidson enrolled at Bryant & Stratton Mercantile College, which had opened a branch of its thriving business school in Buffalo in 1854. Founded in Cleveland in 1853 and embracing the economic optimism of the day, Bryant & Stratton offered ambitious young men the opportunity to systematically develop the skills required for a career in business. Before the advent of such technical schools, apprenticeship followed by employment was the only way to gain a practical business education. The school flourished during the booming 1850s, opening campuses in Albany, Detroit, and Chicago—all growing commercial cities stocked with ambitious young men on the move. Bryant & Stratton offered students detailed instruction in business mathematics, bookkeeping and accounting, and practical penmanship, as well as lectures in ethics, banking, finance, political economy, and commercial law. In 1857, the college charged forty dollars for a ten- to fourteen-week course of study, more than two months of wages for the typical lake sailor of the period and a huge investment for an orphaned teenager.[3]

Based on the content of the textbooks they began publishing by 1860, the Bryant & Stratton curriculum was rigorous, highly quantitative, and largely self-paced: a perfectly tailored combination for the motivated and intelligent independent learner that Davidson clearly was. Davidson's public letters are written in clear and unadorned prose, and the strong quantitative aptitude is evident in all his enterprises—shipbuilding as well as commercial shipping, banking, and sugar beets. Bryant & Stratton did not teach elocution, and Davidson earned a reputation for sometimes being a bit rough in his speech. During his business career, Davidson appeared in court many times, and his recorded testimony indicates a plainspoken man confident in his words and capable of expressing complicated ideas in clear and direct language.

Bryant & Stratton promised more than just instruction; it offered explicit opportunities for hungry young men. The school's advertisements carefully noted that through "the extensive business acquaintance of the Principals, many of the students, on graduating, are placed in lucrative positions." Lest prospective students question just what business connections Bryant & Stratton had, its advertisements prominently displayed a forty-eight-name list of the school's directors. The name of at least one of the directors, General Charles Manning Reed, surely caught Davidson's eye. The "Napoleon of the Lakes" and owner of the *Niagara* and many other fine palace steamers had made a fortune in Great Lakes transportation and was a millionaire several times over.[4] Such advertisements were more than window dressing. One of the Bryant & Stratton strategies for expansion was to get well-known local businessmen invested in the new branch of the school. They did this with invitations to lecture and through positions as college directors. Davidson's consistent inclusion of his formal business education in biographical sketches produced during his lifetime suggests that he attached a strong value to his Bryant & Stratton education.

When Davidson began to learn shipbuilding remains uncertain, and there is no record of his completing a formal apprenticeship—the traditional approach to learning shipbuilding during the wooden age. Whatever his specific circumstances, Davidson began his shipbuilding career during the technological peak of American wooden shipbuilding. Buffalo was home to two cutting-edge shipbuilding firms: Bidwell & Banta, the builders of the *Niagara*, and the yard of B. B. and F. N. Jones, two of the sons of pioneer Ohio shipbuilder Augustus Jones. During the period when Davidson began learning the trade in the 1850s, the two yards built some of the largest and most ornate passenger steamers in the world.

The key point is that Davidson began to learn shipbuilding from some of the most progressive shipbuilders in the country at a pivotal moment in wooden shipbuilding history. He acquired a firsthand knowledge of the latest iron strapping and trussing techniques necessary in building the largest wooden vessels. These engineering lessons were not lost on him. Working at a large shipyard, he also learned about the organization and management of large numbers of men and machinery, the allocation of production space, and materials logistics.

Davidson also absorbed the economic lessons from the palace steamer era. Building a vessel or operating a shipyard out of sync with economic circumstances offered a quick road to bankruptcy. Size, speed, splendor, and innovation could be valuable in a new vessel, but to succeed in business—to get ahead and stay ahead—building and owning the right vessel for the time was far more important.

SALTWATER MATRICULATION

The Panic of 1857 had initiated a protracted downturn in overall commercial maritime opportunities on the Great Lakes, with dozens of lake craft leaving for the ocean between 1858 and 1860.[5] The poor economy meant fewer berths and lower wages for the typical Great Lakes sailor. The next chapter in Davidson's maritime education commenced when he left the Great Lakes for New England in 1860 and found service aboard Atlantic packet and clipper ships.

Economic circumstances probably did not drive Davidson to salt water. A well-established young man with shipbuilding skills and experience as a ship's officer and possibly as a schooner captain, he was well positioned to find work ashore and afloat. More than likely, he succumbed to a young man's dreams of adventure and a desire to see the oceans and the foreign ports he had heard about from countless saltwater mariners who came to the Great Lakes.

If Davidson sought adventure, he surely found it as a sailor for the Black Ball Line of sailing packet ships. Established in 1818, the Black Ball Line revolutionized international transportation by offering the first regularly scheduled trans-Atlantic service. Twice per month, as regular as clockwork, a Black Ball ship would leave on schedule from New York City and Liverpool, England. The Black Ball Line operated what many experts consider the finest wooden sailing vessels ever built. Although the dominance of the Atlantic sailing packet peaked in the late 1840s, high profits in the immigrant trade led to the building of larger and more sophisticated packets into the early 1850s. When Davidson shipped out in late 1860, sailing packet ships on long-distance routes remained profitable.

James Davidson earned his sailing credentials at the toughest training grounds in the Atlantic world. If the sailing world of the mid-nineteenth century had major league teams, then Black Ball packet ship *Great Western* (shown here in 1839) and clipper ship *Panther* would have been the New York Yankees and the Boston Red Sox. SOTHEBY'S

Davidson first shipped aboard the 1,443-ton *Great Western*, launched by William Webb in 1851 for the C. H. Marshall Company (the Black Ball Line). At 191 feet, the ship was not in the same league as the largest Great Lakes steamers produced that year. However, with a breadth of more than 42 feet and depth greater than 28 feet, the *Great Western* had more cargo capacity and sea-keeping ability than any sailing vessel on the Great Lakes. Although the *Great Western*'s design privileged capacity over speed, both virtues contributed to the economic success in the packet ship business. Compared with other deep-water sailing vessels (except clippers), the packet ships were fast, with eastward voyages to England averaging a little more than three weeks. In 1853, the *Great Western* made the run from New York to Liverpool in just sixteen days. The average westward voyage was substantially longer, about thirty-four days, during the 1840s.[6]

The income for a Black Ball Line packet captain came from a share of the vessel revenues and depended upon meeting the line's schedule. Iron-fisted packet captains were among the hardest-driving and most skilled deep-water mariners in Atlantic history. Whatever James Davidson knew about

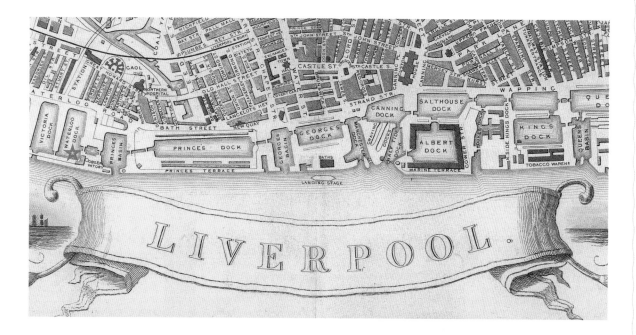

sailing and commanding a ship when he left the Great Lakes was dwarfed by the skills and experience of his first captain, William G. Furber, an old salt who had commanded the *Great Western* since 1852.

The captains commanded hard men and drove them hard as well. Writing in the late 1840s, sailor-author Herman Melville declared that "packet sailors were the toughest class of men in all respects. They could stand the worst weather, food, and usage, and put up with less sleep, more rum and harder knocks than any other sailors."[7] Although the specific conditions on the *Great Western* are unknown, by the late 1850s, shipboard brutality was rife aboard the packet ships. Gangs such as the Liverpool's vicious brotherhood the Bloody Forties terrorized fellow crewmen, passengers, and officers.[8] On the "Old Line," as the Black Ball was known, survival of the fittest was the rule.[9] By simply surviving, Davidson proved his mettle. By any measure, sailing the Atlantic packets in midwinter made him a member of one of the toughest and most highly skilled maritime fraternities in the world.

Packet service also introduced Davidson to Liverpool, then home to the greatest maritime facilities in the world. Davidson grew up in a booming frontier port, but with up to three hundred ships arriving on a single tide in 1847, Liverpool hosted maritime enterprise on an entirely different scale. The *Great Western* moored at the Waterloo Dock, part of an integrated two-hundred-acre dock system built between 1824 and 1866. These

This bird's-eye illustration of Liverpool from 1851 shows many of the city's docks, including the Waterloo, where the *Great Western* docked. DAVID RUMSEY MAP COLLECTION well-engineered and logically arranged docks with their swing bridges and adjacent fireproof warehouses contrasted sharply with the rickety wooden grain elevators that lined Buffalo Creek. It may be no coincidence that the Saginaw River slips of Davidson's later West Bay City shipyard resembled Liverpool's carefully arranged Mersey River harbors.[10]

Leaving the *Great Western*, Davidson signed on to the Boston-owned ship *Panther* in late 1861. Roughly the same size as the *Great Western*, the *Panther* was a true clipper ship, one built to make fast, long-distance passages from the Atlantic to the Indian and Pacific Oceans. Built in 1854 by Paul Curtis in Medford, Massachusetts, the *Panther* was a strong vessel that featured the new iron cross bracing.[11]

On Davidson's voyage to Calcutta, India, a run the clipper had made several times, the *Panther* carried a thousand tons of English railroad iron. Once again, Davidson sailed under a well-tested captain. Born in Hartford, Connecticut, in 1808, John Palsgrave Gannett had commanded large sail and steam vessels for decades before Davidson's voyage and had been the master and part owner of the *Panther* since 1857. The *Panther* remained in Calcutta for nearly two months, giving Davidson plenty of time to explore the most important Indian Ocean port in the British Empire. On April 10, 1861, the *Panther* embarked on what proved an excruciatingly slow 142-day voyage to Boston. After enduring calms in the Bay of Bengal and battling a series of storms off Madagascar, Davidson and the *Panther* arrived in Boston on August 31, 1862. This proved to be Davidson's final saltwater service and Captain Gannett's last voyage on the *Panther*. The ship wrecked off the coast of Vancouver Island with no loss of life on January 17, 1874, and a century later became one of the first British Columbia shipwrecks to be documented by underwater archaeologists.[12]

Davidson remained proud of his saltwater sailing, and occasionally referred to it in public statements made throughout his later life. While we cannot know the specific ways his packet and clipper ship service influenced his later career as a ship captain or shipbuilder, such raw and immersive experiential education was comprehensive and its lessons long lasting. He had sailed some of the world's most treacherous waters and strolled its finest ports. In submitting to the authority of iron-fisted captains, true masters at the art of maritime command, he had extraordinary opportunities to learn how to lead and to discipline the toughest of men. In the *Great Western* and the *Panther*, he experienced firsthand the combination of the speed, seaworthiness, cargo capacity, and durability that made the iron-strapped packets and clippers arguably the finest large wooden sailing ships ever built.

RETURN TO THE GREAT LAKES

After his time on the *Panther*, James Davidson returned to the Great Lakes and secured a government position as an officer aboard the US Lighthouse Service schooner *Watchful*, one of the first dedicated lighthouse tenders on the Great Lakes. At only 147 tons, the *Watchful* must have seemed like a toy after the *Panther* and the *Great Western*, but it was good work and an effective way to stay clear of the Civil War draft. How long Davidson remained in federal service is unclear, but he probably continued on the *Watchful* until late 1864 or early 1865. In January 1865, Davidson married Ellen M. Rogers of Buffalo.[13]

During the 1865 season, Davidson sailed as captain and purchased a half interest in a small old schooner, probably the 114-ton *Seagull* built in Milan, Ohio, in 1846. Successful in 1865, Davidson continued to invest in older schooners, buying and selling them with some frequency. In fall of 1866, he sold the old schooner *Philena Mills* for $6,500. There is no record of his acquiring the vessel, which had capsized in the Saginaw River under a previous owner earlier in the year. Perhaps the skilled Davidson acquired the vessel as salvage and subsequently repaired it. In late 1866, Davidson paid $13,500 for the canal schooner *Lewis Wells*. It is unclear how long he kept that vessel, but possibly only for a short time. In any event, Davidson's economic and social status rose quickly, and in the winter of 1867, Captain Davidson, as the Buffalo City Directory now listed him, moved his wife and newborn son, Edward, into the large brick home at 324 Terrace Street that they would occupy for the next thirteen years.[14]

SHIPBUILDING

In addition to commanding, buying, and selling schooners, Davidson returned to shipbuilding at the Bidwell yard (by then known as Mason & Bidwell) during the winter of 1865–1866. In 1867, he became superintendent at the prolific Bailey Brothers shipyard in Toledo, Ohio, where he maintained an affiliation for the next four years. It was from the Bailey Brothers Shipyard that Davidson launched his first ship in 1870, a 269-ton schooner, the *Laura Belle*, which he named after his baby daughter.[15]

Between shipbuilding work in Toledo and sailing, Davidson probably spent little time at the family residence in Buffalo. In 1871, he rented ground along the Saginaw River near the town of Wenona, Michigan (later West Bay City), and built the *E. M. Davidson* (named after his son Edward), a schooner nearly identical to the *Laura Belle*.[16]

That the first Davidson-built vessels were canal schooners is no surprise. During the booming years of the late 1860s and early 1870s, such vessels might pay for themselves in two years or less. The first Davidson schooners had overall dimensions typical of sailing canallers. The 269-gross-ton *Laura Belle* was 138 feet long, 26 feet in breadth, and 11 feet in depth of hold. The 281-ton *E. M. Davidson* was 142 feet long, 26 feet 2 inches wide, and had a depth of hold of 12 feet. The 1871 Inland Lloyds insurance register valued each schooner at $20,000. Supplement 3 of the register reported the schooners earned an A1 insurance classification (the highest rating) but noted that both had "flat" floors and were equipped with bilge pumps. This was an unusual remark, at least in the 1871 register. The "floor" is the interior bottom of a vessel. In a traditionally shaped schooner of the time, water penetrating the hull or decks collected in the bilge along the vessel's centerline where it would not compromise stability and could be easily pumped out. The bilge pumps mentioned in the insurance listing must have been supplemental to the standard centerline pumps.[17]

Their flat shapes meant that the *Laura Belle* and the *E. M. Davidson* could carry larger cargoes through the Welland Canal than many schooners of the day. While few canal schooners carried more than twenty thousand bushels of wheat, the *Laura Belle* typically carried more than twenty-one thousand and the *E. M. Davidson* more than twenty-two thousand bushels. This one-thousand- to three-thousand-bushel difference Davidson schooners offered translated into substantially higher annual net profits for the owners. In May of 1873, schooners such as the *Laura Belle* and the *E. M. Davidson* could gross nearly $3,000, roughly 15 percent of their construction cost, carrying a single cargo of wheat from Milwaukee to Oswego.

By the early 1870s, the difference between loss or profit and the accumulation of wealth through Great Lakes shipping was becoming a game of inches and fractions of inches. Davidson's ship designs indicate that he understood this fact better than most of his competitors. The ability to haul more cargo than competing vessels of the same class became a hallmark of Davidson-built schooners, schooner-barges, and steamers in the years to come.

As Davidson built the *E. M. Davidson*, the federal government was putting the finishing touches on the St. Clair Flats Ship Canal. From the time of the early palace steamers, the St. Clair Flats near Detroit, Michigan, had been a bottleneck for deeper-draft vessels passing between Lakes Huron and Erie. The new canal was designed in 1869 to provide at least thirteen feet of water and a wider channel.[18] In 1874, work was begun to deepen the channel to sixteen feet.

The new canal and the continued high demand for grain in Buffalo inspired the construction of a new and much larger class of schooner. While the average tonnage of a new schooner built in 1870 included in the 1874 *Board of Lake Underwriters Register* was 243 gross tons, in 1872, it was 408 gross tons, some 68 percent larger.[19] These new schooners approached and, in some instances, exceeded 200 feet in length. Their economic advantages were obvious; the large new vessels required essentially the same number of crewmen and cost about the same to operate as a canal schooner but offered more than twice the cargo-carrying capacity. James Davidson soon followed this trend.

In April 1872, Davidson placed an advertisement in the *Chicago Daily Tribune*: "Wanted—40 ship carpenters to work on a new vessel. Will pay highest going wages. Apply to Captain James Davidson, East Saginaw, Mich."[20] The following month, the *Tribune* offered a brief description of Davidson's new project:

> Captain James Davidson, of Saginaw City, has now under construction the largest vessel ever built in the Saginaw Valley. When finished, she will be a three-masted schooner, with iron rigging, and an extreme length of 205 feet. She is built entirely of first-class oak. Her carrying capacity will be 50,000 bushels of grain, or 1,300 gross tons of ore.[21]

Davidson built his first three ships on different parcels of rented land, the first at the Bailey Brothers Shipyard in Toledo and the next two at convenient places along the banks of Michigan's Saginaw River. In building his first ships, Davidson needed only skilled shipbuilding labor, basic tools, reasonably priced quality timber, and access to sheltered deep water. The projects did not require an established plant, machine tools, or industrial-scale capital. Davidson's son and partner, James E. Davidson, wrote of early schooner construction on the Saginaw River:

> It was customary in those days for a person building a ship to lease a plot of ground on the Saginaw River, build the ship, launch it, and that completed the operation. The ships in those days were built entirely by hand. The timbers were hoisted on high sawhorses and then, with one man standing on top and another beneath the timbers, they were sawn into frames, planking, ceiling and deck beams.[22]

Davidson launched the new schooner, the *Kate Winslow*, on September 15, 1872. At 202.5 feet in length, 34.8 feet wide, and with 13.3 feet depth of hold, the schooner could race easily through the new St. Clair Ship Canal carrying fifty-two thousand bushels of corn. He sold the vessel for $60,000 to H. J. and N. C. Winslow of Cleveland, members of one of the most prominent shipping families on the Great Lakes.[23]

By the early 1870s, three decades of exponential growth in Great Lakes commerce began to exhaust supplies of suitable timber adjacent to the older shipbuilding centers along the western and southern shores of Lake Erie. Transporting timber to places such as Buffalo, Cleveland, and Toledo added significantly to the cost of building a ship. The demand for quality timber during flush economic times could push shipbuilding costs up even further. Davidson addressed these challenges by locating his yard adjacent to some of Michigan's finest stands of white pine. The huge pines, up to 175 feet tall and seven feet in diameter, were not the only trees with commercial value. A large variety of soft and hard woods also abounded, including oak—the shipbuilder's material of choice.[24] These forests had attracted the attention of eastern lumbermen in the 1850s. In 1863, lumber baron Henry Sage purchased 116 acres of land on the west side of the Saginaw River and within two years put into operation one of the largest, most modern, and most efficient sawmills in the Great Lakes region. Sage and his partners also invested nearly a million dollars into the carefully planned company town of Wenona—where Davidson bought land and established his first yard with the profits from the sale of the *Kate Winslow* and his other ventures.[25]

The *Kate Winslow*, shown here carrying lumber, was built in 1872 to carry grain. It wrecked in northern Lake Michigan carrying 1,200 tons of iron ore in October, 1897. C. PATRICK LABADIE COLLECTION / THUNDER BAY NATIONAL MARINE SANCTUARY, ALPENA, MI

Davidson was not the first shipbuilder to recognize the potential for shipbuilding of the Saginaw River Valley. William Crosthwaite, who began his shipbuilding career at Bidwell & Banta in 1841, established a yard at Bay City in the mid-1860s, building a dry-dock and completing a few large vessels before moving his operation to East Saginaw. Chaumont, New York, shipbuilder Chelsey Wheeler moved his family and opened a yard at Saginaw in 1866. In 1875, Wheeler's son Frank would establish a major shipyard at Bay City and become Davidson's principal competitor and professional nemesis. By 1875, more than 112 manufacturing businesses reflecting dozens of industries, including thirty-eight sawmills and four shipyards, operated at or adjacent to Bay City.[26]

THE WOODEN SHIP IN THE INDUSTRIAL AGE

Davidson invested his shipbuilding and ship-owning profits into his own shipyard. In 1873, he became one of the earliest Great Lakes shipbuilders to integrate a steam-powered saw into ship construction. Although pioneered by the famous East Boston shipbuilder Donald McKay in the 1850s, steam-powered saws remained uncommon in the 1870s shipyards and illustrate Davidson's progressive mentality. The new Muley sawmill's vertical blade could make up to three hundred board cuts per minute, potentially speeding up construction and reducing labor costs. The original saw, however, wasted too much wood, and Davidson did not replace it after its destruction by fire in 1874.[27]

Davidson had begun his shipbuilding career by producing large grain schooners. While considered modern for the time, these were up-to-date variants of a well-established form of merchant vessel that sailed coastal and Great Lakes waters throughout the nineteenth century. Davidson's fourth ship, the *James Davidson*, a steam-powered bulk cargo carrier of the latest design and largest class, was a different matter. Responding to the same economies of scale inspiring the construction of the large schooners, the new bulk carriers were harbingers of an industrial age. Huge for their time, the early bulk steamers were small-scale versions of what Great Lakes cargo ships would become. Rather than larger and more efficient versions of the classic sailing ships, these were the beginning of something new.

True "revolutionary" ships are rare; however, marine historians generally agree that the bulk carrier *R. J. Hackett*, designed by Captain Elihu Peck, sparked a revolution that transformed Great Lakes ships and shipping. The *R. J. Hackett*, launched in Cleveland, Ohio, in 1869, became the prototype for the modern Great Lakes bulk carrier. With the engineering spaces in the stern and a raised pilothouse pushed forward to the bow, at first glance the *R. J. Hackett* resembled the smaller "steam-barges" that found favor in the lumber trade during the 1860s. At 210 feet in length, with a displacement of 1,192 tons and a powerful engine, the ship's dimensions and engineering features were more in line with the larger screw-propelled passenger and package freight steamers of the period than with the sailing vessels that carried most of the bulk commodities.

Functionality and efficiency made the design of the *R. J. Hackett* exceptional. The pilothouse perched at the extreme forward end of the ship provided superior visibility, allowing the officers to better avoid collision or grounding when maneuvering in tight channels and congested harbors or navigation lanes. Placing the propulsion machinery near the stern eliminated excessively long, expensive, and vulnerable propeller shafts. Although technically a "double-decker," the ship had one large open hold.

The design of the *R. J. Hackett* was not just about the safe carriage of large cargoes; its deck hatches were spaced to mate perfectly with the gravity-fed loading chutes of the new Marquette iron ore docks and enabled the rapid loading of bulk cargo. Instead of ceiling planking parallel to the keelson, as on a schooner, the *R. J. Hackett*'s ran perpendicular. Sitting flat atop a series of longitudinal keelsons, the ceiling planking provided a uniform surface that proved easier to unload and to repair. Equipped with a powerful engine, the *R. J. Hackett* could tow one or more large unpowered barges or large schooners. This consort system had become increasingly common during the 1860s, but the new bulk carriers took it to new levels of efficiency. Many of the basic features of the *R. J. Hackett* became the norm for new Great Lakes bulk freighters for the next hundred years, and the classic bulk carrier remains a common sight throughout the region today.[28]

Great Lakes shipyards launched at least forty-seven of the double-decked bulk carriers between 1869 and 1874, among them the mammoth *James Davidson*, completed in June 1874. At 1,464 tons and 230.5 feet in length, 37 feet in breadth, and 19.5 feet in depth, the *James Davidson* was one of the largest bulk carriers on the lakes when launched.[29] Captain Davidson may have intended his new steamer to be the largest bulk carrier on the Great Lakes, but other builders beat him to the punch. Two months before the *James Davidson* was completed, David Lester, a Marine City, Michigan, shipbuilder, finished the *V. H. Ketchum*, a bulk carrier 2.5 feet longer, more than 3 feet wider and deeper, and 200 gross tons larger than the *James Davidson*. Originally intended as giant schooner, the *V. H. Ketchum* also

featured a clipper-style bow more pleasing to the nautical eye than the bow of the *James Davidson*. The Quayle and Murphy bulk freighter *Persian*, launched in July 1874 at Cleveland, was longer yet at 243.7 feet.[30]

Both Davidson and Lester saw the value of the larger and more powerful bulk carriers and understood the need for building strong hulls of limited draft, but they took very different approaches in their hull design and engine choices. Coming to grips with this early lack of standardization is essential to understanding the evolution of wooden age Great Lakes bulk carriers.

Comparing the "monsters" of 1874 provides some insight into Davidson's design priorities and illustrates the frontier nature of industrial shipbuilding on the Great Lakes during the period. Although more than 13 percent larger in gross tonnage, the *V. H. Ketchum* had only a slight advantage over the *James Davidson* in functional cargo capacity. The largest corn cargo found for the *Ketchum* was sixty-eight thousand bushels unloaded in Buffalo in October 1876. Two months earlier, the *James Davidson* delivered sixty-seven thousand bushels. In terms of efficiency, the *James Davidson* carried 47.8 bushels of corn per gross vessel ton, almost 17 percent more than the *V. H. Ketchum*. The designer of the *V. H. Ketchum* opted for two high-pressure engines as opposed to the single-cylinder low-pressure engine selected by Davidson. Two engines probably increased building costs and added considerable weight in machinery that would have come at the cost of cargo capacity.[31]

The pioneering bulk steamer *R. J. Hackett*, built in Cleveland in 1869, was originally fitted with three masts and auxiliary sails. Modernized multiple times, the *R. J. Hackett* remained in service until claimed by fire in 1905. BOWLING GREEN STATE UNIVERSITY

The *James Davidson* at Duluth, Minnesota. Note the massive arches that were added in the 1877 refit and the small but ornate wheelhouse. THUNDER BAY RESEARCH COLLECTION

No contemporary plans for the *James Davidson*'s hull survive; indeed, it is likely they never existed. An undated and slightly blurry photograph of the *James Davidson* reveals a small, highly stylized wheelhouse and large external bishop arches added in an 1877 refit that swept several feet above the deck at midships. These features gave the ship aesthetic and architectural characteristics associated more with large mid-nineteenth-century passenger steamers than with industrial bulk carriers. As originally configured, the steamer had three masts, including a 106-foot main and an 86-foot mizzen, with large fore and aft sails to provide additional or auxiliary power.[32]

The *James Davidson* wrecked in Lake Huron's Thunder Bay in 1883. The well-preserved remains, protected as part of the Thunder Bay National Marine Sanctuary, provide valuable evidence about the vessel's design. The hull's longitudinal strength came from a heavy keelson assembly with a large central keelson and two sister keelsons. Outboard from each side of the keelson assembly are four bilge keelsons of somewhat smaller dimensions. Regularly spaced pairs of heavy frames run underneath the keelsons. Also preserved are remnants of the bishop arches added in 1877.

This intact section of the steamer *James Davidson*'s bilge preserves important information about the first generation of steam bulk carriers on the Great Lakes. Note the lack of visible iron cross bracing that became standard features of later Davidson-built ships. NOAA, THUNDER BAY NATIONAL MARINE SANCTUARY

Two divers studying the remains of the *James Davidson* in the Thunder Bay National Marine Sanctury in Lake Huron. The diver in the background hovers over a section of heavy arch added in 1877. NOAA, THUNDER BAY NATIONAL MARINE SANCTUARY

Propulsion came from a low-pressure steam engine with one 50-inch cylinder having a 40-inch stroke, built in Detroit by the innovative Samuel F. Hodge. Trained as a machinist by his father at the Great Consul mine in Devon, England (at the time the most productive copper mine in the world), Hodge and his young family immigrated to the United States in 1849. Settling in Detroit, he began building equipment for Lake Superior copper mines. He gained several patents for improvements to steam and industrial process machinery and became one of the most respected builders of marine and large, stationary steam machinery in the Great Lakes region. Although a veteran fabricator of steam machinery in 1874, Hodge expanded his fortunes rapidly after the introduction of the industrial bulk carrier, with his company producing more than one hundred large marine steam engines in the 1880s and 1890s. These powerful engines enabled bulk carriers to tow unpowered consort vessels packed with cargo. This consort system proved essential to the economic successes of the early wooden steam bulk carriers. Hodge died in 1883, but the company continued under the leadership of his son, Harry S. Hodge.[33]

From the beginning, Davidson cultivated a public reputation for both style and innovation in his ships. Reporting the launch of the *James Davidson*, one newspaper commented on the ship's "commodious and nicely furnished room abaft of the wheelhouse . . . and roomy and neat" forecastle accommodations as well as a tasteful "white Oakland" grained dining room for the officers and engineers located on the aft end of the ship. More style and innovation came together outside of public view with the Chalmers-Spence asbestos "non-conducting felting" that covered the boilers. The Chalmers-Spence Company was one of the first in the United States to produce asbestos for industrial uses. Applying asbestos to marine boilers was a new idea in 1874, and marine reporters noted its installation in the *James Davidson*. By covering the boiler in fireproof asbestos—material that also resists the conduction of heat—Davidson added a new measure of safety and improved fuel efficiency. The asbestos material was easily sculpted during application, and the *James Davidson* boiler was painted lavender and "literally frescoed" with stripes and gold and red ornamentation.[34]

With the grim economic times, many knowledgeable people around the Great Lakes were skeptical about the giant bulk carriers. The *Detroit Free Press* predicted that given the current business conditions, the *V. H. Ketchum* would become "an elephant of no ordinary dimensions in the hands of her owners." Concerning the new *James Davidson*, as a nautical reporter for the *Oswego Palladium* wryly commented, "Captain Jim, we had given you credit for more sense."[35] Davidson gambled and his ideas ultimately paid off, but only after coping with some challenging seasons.

DEPRESSION AND RECOVERY

By the time James Davidson launched his new steamer in June 1874, the bottom had fallen out of the shipping and shipbuilding businesses. In 1873, vessel owners collected more than eleven cents for the average bushel of wheat carried from Chicago to Buffalo. By 1876, they were lucky to get the same charter at three cents per bushel.[36] Lake traffic fell, with annual overall tonnage shipped declining in 1874, 1875, and 1876.[37] Shipyards across the region scaled back or closed entirely, and many shipbuilders with overextended credit went broke. Davidson suspended work at his Michigan shipyard. But rather than taking a loss on his prize steamer, he elected to operate the steamer himself as owner, manager, and captain.

Davidson launched the steamer in mid-June 1874. The ship was towed to Detroit for the installation of the propulsion machinery and then steamed home to Bay City, ready for its maiden voyage. Captain James Davidson planned to make a big splash with the first trip, boasting that he would carry a record cargo of eleven thousand barrels of salt from Bay City to Chicago.

Things did not go well. The steamer grounded in the Saginaw River after leaving the dock with only six thousand barrels on board. Although the vessel was soon released and additional salt loaded from lighters (smaller boats or barges used to carry cargo out to ships in deeper water), the well-publicized event proved an embarrassing start for the proud captain. Problems continued on the return voyage. The steamer left Chicago bound for Buffalo with a cargo of sixty thousand bushels of wheat but ran hard aground when passing through the treacherous Lake St. Clair. Although no damage to ship or cargo resulted, J. W. Hall, the acerbic marine reporter for the *Chicago Inter Ocean*, termed it "one of the most momentous events of the season."[38]

The *James Davidson* was freed on September 13 and steamed on with more trouble following in its wake. Unusually low water levels were plaguing vessels across the lakes, and while approaching the Buffalo harbor at about nine o'clock on the night of September 15, the steamer grounded hard. After more than a day of effort, it was once again released completely undamaged, except in reputation. On October 10, the ill luck plaguing the *James Davidson* seemed to have departed, and the steamer took on 110,000 bushels of oats, at the time the largest cargo of that commodity ever carried on the Great Lakes. The good luck did not last. In November, the ship ran aground near Grosse Point in Lake St. Clair and later in Lake Huron, sustaining no damage in either incident.[39]

November may have been the worst month in Davidson's early career as a captain. The two groundings were minor events compared to repeated incidents in Chicago that made the steamer *James Davidson* and its captain the laughingstock of the western Great Lakes.

Chicago, as measured by the number of ships entering and leaving its harbor, was one of the busiest ports in the world in 1874. That year, the *Chicago Collector* recorded the arrival of nearly eleven thousand vessels, with the majority of the traffic concentrated between the months of May and October. On a typical day, between forty and sixty commercial vessels entered the narrow and shallow confines of the Chicago River, where they could snake several miles into the private wharves in the center of the city. Although generating great profits, the city of Chicago had an uneasy relationship with its port. Each time Davidson brought his giant into Chicago, he had to contend with several of the twenty-seven swing bridges that carried the city's burgeoning traffic. Traffic backups and delays connected with the bridges were an unpleasant fact of life for nineteenth-century Chicagoans.[40]

On Wednesday morning, November 9, 1874, the *James Davidson* entered the river carrying a large cargo of salt. After safely negotiating the main branch of the river, the ship made a sharp turn to port to enter the south branch. Safely passing the Lake Street, Randolph Street, and Washington Street bridges, Captain Davidson signaled the tender of the Madison Street Bridge. The tender stopped the busy morning street traffic and opened the bridge. Following the channel to the west of the bridge pier, the steamer shuddered to a stop, blocking one of the city's busiest thoroughfares. Captain Davidson removed three hundred barrels of salt and backed his lightened steamer into deeper water, but the incident tied up traffic for more than three hours and was, according to the *Chicago Tribune*, the worst such incident in the memory of the bridge tender. The bridge tender laid no fault on the *James Davidson*, which was drawing only 13 feet of water, or on the captain. As had been the case in Buffalo, adverse winds had pushed Lake Michigan water to the west, quickly lowering the water level in the Chicago River some ten inches. Despite the natural cause, a *Tribune* reporter could not resist listing the ship's many recent incidents and declaring, "There is no doubt, however, that the mammoth propeller *James Davidson* is A BIG FAILURE."[41]

With its narrow passages, shallow water, and bridges, the Chicago harbor became a nightmare for mariners and commuters with the appearance of massive new freighters such as the *James Davidson* in early 1870s. CHICAGO HISTORY MUSEUM

The following week, a long article in the *Chicago Tribune* began, "That old stick-in-the-mud, the propeller *James Davidson*, whose huge bulk a week since put an end to communication between East and West Madison streets for three hours and a half, played a similar prank at the same place yesterday evening, at twenty-four minutes past 6 o'clock, when she again succeeded in running aground at exactly the same spot she stirred the mud up in last week." Although most of the article consisted of a humorous description of angry streetcar commuters anxious to get home for supper, the undertone is a not-so-veiled criticism of "the elongated wash-tub" and its captain.[42]

Despite the bad luck and public ridicule, James Davidson and his monster steamer made it through their maiden season intact. Given that Davidson had no previous history commanding a steam vessel, much less one of the newest and largest on Great Lakes, it seems remarkable that things went as well as they did. With many of the bugs worked out of the steamer and Captain Davidson having more experience, the following year went more smoothly. The continued deepening of strategic points on the Great Lakes helped as well by providing those extra inches of water depth that the larger ships needed.

Over the next few years, Davidson ran the steamer hard. With one or more unpowered schooner barges such as the *Ogarita* and the *J. C. King* in tow, Captain Davidson could transport one hundred thousand or more bushels of corn in a single trip. The steamer also commonly carried oats, wheat, coal, and sometimes iron ore or timber. The use of the schooner barges added flexibility. For example, on one trip in August 1878, the *James Davidson* entered Chicago with a mixed cargo of lumber and salt, with the iron ore-laden *Ogarita* in tow. During most of the trip, the *James Davidson* also towed the schooner barge *J. C. King*, which split off at Milwaukee to deliver coal.[43]

Narrow passages, poor harbors, areas of shoal water, high traffic levels, spring and winter ice, and the occasional violent storm all made accidents an inevitable part of the commercial shipping business. Throughout the second half of the nineteenth century, the Great Lakes had the worst safety statistics of any maritime region in the United States. The advent of larger bulk freight steamers and the increasing use of tow barges in the 1870s made matters even more difficult and resulted in much wear and tear on ships and men. In 1875, after less than a full year of service, Captain Davidson had Buffalo's Union Dry Dock Company replace the steamer's 12-foot-diameter propeller. Two years later, in midseason, he brought the ship back to the Union Dry Dock for a month-long overhaul that included refastening the planking, installing new anchors, and—most significant—installing large bishop arches. The arches added substantial longitudinal strength to the vessel and suggest that, as originally constructed, the *James Davidson*'s long hull flexed too much under certain conditions. This may explain the need to refasten the hull of such a young vessel. The modifications did not affect the steamer's cargo capacity. On the first trip back in service, the *James Davidson* loaded 105,000 bushels of oats at Chicago.[44]

The *James Davidson*'s profitability during this period can be only roughly estimated. Freight rates and the price of labor fluctuated greatly depending on market conditions. Often these were tied to the seasons of the year. Early and especially late-season cargoes typically fetched the highest rates. Late-season voyages often required paying premium wages to the crews and resulted in extra wear and tear on the vessel.

An early October 1877 charter to carry coal from Buffalo to Chicago garnered thirty cents per ton, or between $400 and $450 in revenue for the steamer. The receipts for its unpowered consort, *J. C. King*, contributed an estimated $150. At the end of October, the *James Davidson* and the *J. C. King* took on another load of coal for Chicago. Although the terms were kept private, other vessels were getting seventy-five cents per ton, a rate that would have yielded an estimated trip revenue of $1,125.[45]

Late-season navigation imposed higher risks. On November 1, 1877, the *James Davidson* left Buffalo towing the *J. C. King*. Ten days later, both vessels limped into Chicago, their sails shredded and the *James Davidson*'s gaff and booms torn away—the victims of a vicious Lake Huron storm that sank one vessel and seriously damaged many others. On November 18, the *James Davidson* cleared Chicago for Buffalo carrying fifty-seven thousand bushels of wheat. On November 26, the *James Davidson* left Buffalo with another cargo of coal for Chicago. The freight rates for these trips are unknown, but the wheat probably generated between $2,000 and $2,500, and the coal, about $1,500 in freight receipts.[46]

The annual expense of operating the *James Davidson* is unknown. Serving as the vessel's captain saved Davidson at least $1,000 per season, possibly more. A key feature of Davidson's business strategy was to keep his vessels working whenever possible, even when rates were low. The *James Davidson* was always one of the first steamers to go into service in the spring and among the last to go into winter quarters. It was a strategy Captain Davidson retained throughout his long career.

Davidson remained in command of the steamer from the summer of 1874 through most of the 1879 season. During that time, he faced all of the typical dangers the Great Lakes offered, among them an 1879 collision that sank his schooner barge *Ogarita* in Canadian waters near Detroit. Davidson raised and thoroughly rebuilt the *Ogarita* and eventually won a major lawsuit connected with the accident. This was probably not Davidson's first venture into court and would not be his last. Davidson had begun to learn about commercial law as a young man at Bryant & Stratton, and over time, he seems to have mastered all of the areas of law connected with his shipping and business affairs. Relentless, methodical, well-funded, and confident, Davidson rarely lost in court. He collected $12,124 from the owners of the offending vessel, a figure more than the estimated value of the *Ogarita* at the time of the accident.[47]

With the return of good times, Davidson's careful amassing of capital and herculean efforts as a captain and manager began to pay off. Late in the 1879 season, he relinquished command of the *James Davidson* and returned to his shipbuilding on the Saginaw River.

In March 1880, he purchased the Ballentine & Company Shipyard managed by F. W. Wheeler, transferring the equipment to his old shipyard property in West Bay City where he was building a 300-foot dry-dock large enough to accommodate any vessel then working the Great Lakes.[48] Davidson's knowledge of business practice, years of experience operating his giant steamer, knowledge of shipbuilding, and ownership of Saginaw River properties placed him in a perfect position to prosper as the Great Lakes and national economies recovered. He understood that achieving long-term success in Great Lakes shipping required mastering the challenge of building and maintaining cost-efficient vessels capable of turning a profit even when shipping rates were low.

Most of the region's major ship owners and builders, including Davidson, recognized that the future lay in larger ships. However, this group was far from unanimous in believing that the best answer was to build a larger class of *R. J. Hackett*–style bulk carriers.

In 1881, James Davidson's former employers, the Bailey Brothers, launched the United States' first five-masted merchant schooner. At 265 feet in length, 37 feet in breadth, 18 feet in depth of hold, with masts stretching 150 feet into the sky, the *David Dows* was magnificent. However, built at a contracted price of $78,000, completing the *David Dows* bankrupted the Bailey Brothers Shipyard.[49] In 1883, after a multiyear struggle, Ohio builder Valentine Fries launched the giant 278-foot, four-masted schooner *Golden Age*. These giant sailing vessels had the largest cargo capacities on the lakes but proved awkward to handle in port and possessed all of the limitations and vulnerabilities of smaller wind-dependent vessels. The *Golden Age* soon ended up being towed behind bulk carriers as a giant schooner barge and survived into the 1920s.[50] The *David Dows* had less luck. Its rig reduced to a schooner barge in 1883, the ship sank near Chicago in 1889, the victim of a winter storm.

On the other end of the technology spectrum, in 1882, the Globe Ship Building Company and a consortium of Cleveland investors launched the *Onoko*, the Great Lakes' first iron-hulled bulk carrier. Stronger than wood, iron allowed for the construction of large vessels much lighter in weight than wooden ships of comparable dimensions. The lighter weight translated into more cargo capacity. The *Onoko* was the longest bulk carrier on the lakes at 287 feet in length when launched, and it possessed a working cargo capacity that remained unmatched by any other Great Lakes ship for many years. Although ultimately judged an immense success, the *Onoko* was in some ways a risky investment. It carried only 25 to 30 percent more cargo than the largest class of wooden bulk carrier of 1882 but cost nearly twice as much— some $210,000. As shipbuilding revived in the early 1880s, building a wooden ship on the *R. J. Hackett* model equipped with the most modern engine and other machinery represented a safer bet and a better short- to medium-term investment.[51] This is the direction Davidson took.

By 1880, the early advantages of the Saginaw River region for the builders of wooden ships had become more pronounced. Older shipbuilding centers of southern and eastern Lake Erie relied on timber imported from increasingly long distances. In addition, in large cities such as Buffalo and Cleveland, the wages for skilled and unskilled labor was higher. With access to quality labor and lumber at low prices, James Davidson was perfectly located to build and service a new generation of giant wooden bulk carriers.

Captain Davidson spent 1880 organizing the revived shipyard and managing the operations of the *James Davidson*. At the close of the 1880 season, the *James Davidson* suffered serious damage after striking ice when en route to Buffalo with sixty thousand bushels of corn. The ship limped into Marine City, Michigan, with three feet of water in the hold and much damaged cargo.[52]

Through it all, Davidson was working on the design for a new steamer, the powerful *Oceanica*, whose construction began in October 1880. Davidson financed the project himself, selling the schooner barge *Ogarita* for $28,000 and the *James Davidson* for a reported $75,000. Davidson's other liabilities, assets, and sources of income are unknown, but the $103,000 he received for the two well-used vessels likely covered the full cost of the new *Oceanica*.

James Davidson's shipyard. COURTESY OF THE WHS MARITIME PRESERVATION AND ARCHAEOLOGY PROGRAM

Designed for Great Lakes and Atlantic Ocean service, the *Oceanica* when launched in 1881 placed James Davidson with the first rank of Great Lakes shipbuilders. The ship survived nearly four decades of hard service before burning on the St. Lawrence River in 1919. BOWLING GREEN STATE UNIVERSITY

Although showing signs of encouragement on the Great Lakes, the general condition of American shipping and shipbuilding was of national concern in 1880. From leading the world in shipbuilding in the early 1850s and rivaling Great Britain in merchant shipping, the United States' maritime industries had become an international embarrassment. A reflection of such concerns was the investigation by Henry Hall, a special agent for the federal government who, in November 1880, began fieldwork for a landmark study of American shipbuilding. Hall personally inspected and collected data from shipyards on the Atlantic and Pacific coasts, the Mississippi River, and the Great Lakes. Hall developed a comprehensive understanding of American ships, shipbuilding techniques, and shipbuilders probably unmatched in the history of the United States.[53]

Hall visited Bay City, Michigan, on August 25 and 26, 1881. Taking copious notes, Hall captured the creativity and spirit of enterprise in the air and described the Saginaw River shipyards as the "finest ship-yards for wooden ship-building in the United States." He was especially impressed by the "admirable manner in which steam has been utilized for as many purposes as possible and the thorough system employed in all branches of the work."[54]

The final report published by the US Census Bureau in 1884 provides a detailed description of an unnamed shipyard that could fit either the James Davidson or the Frank Wheeler facility:

A railroad runs past the yard, by means of which a great deal of the heavy oak is brought to the spot. The logs are rolled from the platform cars and lie upon the ground until called for, and a long rope from within the shops is attached to each log as it lies on the ground, or the end of the trunk is lifted up and supported by a pair of wheels and the rope is then attached. Steam-power is then applied, and the log is drawn across the railroad track down to the yard, through the gate, and along the beaten path to the saw-mill. It is there cut into flitch or square timber, as the case may be, planed, and passed on, all the while by steam-power, to a place in the yard where it can lie and season. The stick never changes end, and no steps are wasted in the handling of it. Whether designed to go through the bevel saw for fashioning into a frame or intended for the steam box to be softened for planking it passes on and on until it is finally lifted to its place in the vessel by a steam derrick. The yard has a complete equipment of bevel, jig, and joiner's saws, bolt-cutters, steam boxes, planers, treenail machines, and small tools. These arrangements were originally made necessary by the scarcity of labor; but they especially fulfill the purposes of saving expense in constructing the ship, and it is claimed that one-half of the labor formally required is now saved.[55]

Davidson and Wheeler each nurtured high expectations for the increased Great Lakes–ocean trade projected for the anticipated opening of the larger Canadian locks on the St. Lawrence River.[56] When Hall visited, both men were completing very strong iron-strapped and banded wooden bulk steamers. These were large vessels for the time, yet just small enough to transit through the projected new locks to the St. Lawrence River and thence on to the Atlantic Ocean. Wheeler's *Clyde* and Davidson's *Oceanica* proved durable and productive vessels, and both survived more than four decades of hard service.

Externally, the new steamers retained the functional shape and deck layout pioneered in the *R. J. Hackett*: a high pilothouse pushed forward to the bow, a large open hold with standardized hatches, a powerful engine, and additional quarters located near the stern. With a keel length of 245 feet and overall length of 256 feet, Wheeler's *Clyde* was an impressive sight. The 1,306-gross-ton vessel had a 36.6-foot beam and a 19.3-foot depth of hold. The vessel's four tall masts with gaff-rigged booms and auxiliary sails increased the visual sense of the steamer's large size.

The *Clyde*'s small and rather ornate pilothouse, similar to the *James Davidson* of 1874, however, gave the vessel a somewhat antiquated look.

This may have reflected the tastes of the *Clyde*'s actual builder, the venerable Captain Frederick Nelson (F. N.) Jones of the famous Jones family. F. N. Jones's career spanned the technological history of the wooden age on the Great Lakes. The fourth son of Augustus Jones, F. N. Jones was born in Essex, Connecticut, in 1815 and began to learn shipbuilding at his father's yard on the Ohio frontier during the 1820s. At age twenty-two, he built his first passenger steamer, the famous *Bunker Hill*, in 1837. An aspiring entrepreneur, at age twenty-three he built and became the owner and captain of the luxurious sidewheel steamer *Lexington*. Jones was an innovator throughout much of his early career, building the new propellers of the 1840s and some of the largest palace steamers of the 1850s. During the 1860s and 1870s, Jones built a few larger schooners and several small steam tugs at his Tonawanda, New York, shipyard. Like his father Augustus and many of his contemporaries, however, F. N. Jones was a better shipbuilder than businessman, and the depression years of the 1870s weighed hard on him. In public testimony before a legislative committee studying unfair business practices by the railroad in September 1879, Jones (who also owned at least one schooner) reported that he had made little money during the previous four years.[57] This probably explains Jones's taking the position working for Frank Wheeler at sixty-five years old, an advanced age for a nineteenth-century shipbuilder.

Lacking experience in the building of large, complex steamers, young Frank Wheeler had wisely brought F. N. Jones to Bay City to superintend the building of the 1,600-ton package/passenger steamers *Lycoming* and *Conemaugh*, launched in 1880. The *Clyde* was F. N. Jones's third and final ship at the Wheeler yard, and his only "modern" industrial bulk steamer. After completing the job, he returned home to Buffalo, where he died less than two years later.

The *Clyde* was a fine vessel, and inspector Henry Hall recorded two sentences in his notebook giving it good marks. By contrast, however, he devoted over a page of closely detailed notes to Davidson's *Oceanica* and provided a detailed account in his published report (sections of which are abridged below):[58]

> She measures 1,490 tons, was built for ocean service, if required, and is 250 feet keel, 265½ feet in length over all, 38¾ feet beam, 19½ feet hold, with two decks and three masts. . . . The frames were cut from 6-inch flitch, and the floors were trebled to 4 feet up the bilge; they were molded 16 and sided 18 inches over the keel, and spaced 22 inches. The

main keelson was 16 by 17 inches; the two sister keelsons were 14 by 13 inches. Six floor keelsons on each side of the main keelson were 13 by 12 inches, all securely bolted through the frames with 1⅛ and 1½ inch iron.

The principal change between the largest wooden bulk carriers of the early 1870s and those of the early 1880s was the substitution of high external bishop arches with complex systems of iron cross-bracing, banding, trusses, and integrated arches in line with those previously used on the largest palace steamers, clipper ships, and packet ships during the late 1840s and 1850s. Davidson could have incorporated iron trussing and strapping when he built the *James Davidson* in 1874. He had a direct knowledge of features and had tried to avoid using the bishop arch method of strengthening the hull. It seems likely the cost of the iron, additional tools, and labor required for this method was too high for those times. Although iron cross-bracing was becoming the norm in the few large steamers built in the late 1870s, the shift away from the bishop arch and to fully integrated iron basket trussing in bulk carriers seems to have occurred abruptly in about 1880. That year, Cleveland's well-respected Thomas Quayle and Sons shipyard launched the *Wocoken* and the *Henry Chisholm*, bulk carriers with keel lengths of about 250 feet built with the iron basket-style architecture. Other builders, including Davidson and Wheeler, followed suit. In any event, from 1880, iron and later steel cross-bracing, bands, and arches became the standard in large wooden bulk carriers on the Great Lakes.

Hall provided a concise description of the principal iron features incorporated into the *Oceanica*'s hull:

> The main strap, running around the vessel near the plank-shear, is 10 inches wide and ⅞ inch thick, which the diagonal straps are 5 inches wide and ½ inch thick, and are riveted to the main strap and bolted through frames and ceiling. Amidships a few of the diagonal straps cross each other at right angles and are riveted to each other.[59]

Davidson launched the *Oceanica* at ten past three on the afternoon of July 30, 1881. The Saturday afternoon spectacle drew an estimated five thousand people to the banks of the Saginaw River.[60] During the following seven weeks, engineers installed the ship's machinery. To power the *Oceanica*,

The innovative ship-building career of Captain Frederick Nelson Jones (1815–1883) spanned seven decades. Captain Jones's last ship was the modern steam bulk freighter *Clyde*, launched in Michigan in 1881. JIM JONES

Davidson again turned to Samuel Hodge of Detroit. Unlike the *James David-son*, which was powered by a single-cylinder engine, the *Oceanica* was equipped with a two-cylinder compound engine. Steam from the boiler went first to power a 30-inch-diameter high-pressure cylinder, then passed into a second 50-inch low-pressure cylinder. The compound engine offered more power, greater efficiency, and some redundancy, as the ship could be configured to run on the high-pressure cylinder alone, if necessary. Other improvements in technology included an independent steam-powered cargo hoisting engine, a first-class Providence anchor windlass, and a steam-powered reversing gear to aid with ship handling.[61]

On September 1, 1881, the *Oceanica* steamed into Chicago with a record thirteen thousand barrels of Bay County salt. The *Chicago Tribune* concluded a glowing description of "one of the finest boats on fresh water" by stating, "The *Oceanica* is not a steam-barge or propeller, but a veritable steamship," a distinction repeatedly made by its builder. By the summer of 1881, Davidson was planning an even larger steamer, and he sold the *Oceanica* to the Lehigh Valley Transit line shortly after its completion for a reported $118,000.[62]

Davidson's next steamer was modeled after the *Oceanica* but incorporated some changes. At 272 feet long and 1,609 gross tons, the *Siberia* was 6½ feet longer and more than 100 gross tons larger, a few inches wider, and a critical 1½ feet shallower in depth than the *Oceanica*—just 18 feet. To accommodate the additional length and compensate for the reduced depth, Davidson increased the size of the ⅞-inch-thick iron belt surrounding the hull from the 10 inches used on the *Oceanica* to 16 inches. The new ship also had an 8-inch-wide by ¾-inch-thick iron arch running forward from the fantail (stern). This additional iron may have allowed Davidson to reduce the sided dimensions (width) of the main keelson from 17 to 16 inches. Other minor changes likely occurred in the hull architecture as well. On deck, the ship had one additional hatch for a total of six, three masts, and the same basic deck machinery as the *Oceanica*. Rather than the Hodge Works at Detroit, the *Siberia*'s compound engine was fabricated in Buffalo by Henry Trout at the King Iron Works. Sources conflict, but it appears likely that the *Siberia* had a more powerful engine than *Oceanica*.[63]

Davidson built the *Siberia* for himself and took command for the ship's maiden voyage on August 1, 1881. Approaching fog-bound Chicago, the *Siberia* collided with the heavy crib of a new and unmarked breakwater. The *Siberia* crashed headlong into the crib, destroying several of its timbers. The *Chicago Inter Ocean* reporter remarked, "How she could strike such a solid,

The *Siberia* loading coal at Fort Williams, Ontario. The three masts visible in the photograph were removed in later years to improve efficiency. C. PATRICK LABADIE COLLECTION / THUNDER BAY NATIONAL MARINE SANCTUARY, ALPENA, MI

square blow WITHOUT SPRINGING HER HULL is most wonderful. She is certainly a most strongly built craft." Fortunately, Davidson was not trying to set any records and had loaded only a modest 9,812 barrels of salt aboard the *Siberia*. Three weeks later, the *Inter Ocean* lauded the ship's superior handling: "This monster craft, without any tug convoying her, laden with 2,100 tons of coal, went all the way through our narrow, crooked harbor up to the dock at North Avenue Bridge without a foul, and without even grazing any other craft or bridge."[64]

Davidson commanded the *Siberia* for the first two round trips between Chicago and Buffalo. The first trip went well. With two barges in tow, the *Siberia* covered the distance from Chicago to Buffalo in five days. Drawing 14 feet 7 inches of water, the ship delivered 74,040 bushels of wheat (2,291 tons).[65] On September 2, Davidson showed his old desire for public spectacle and announced his intention to have the fully loaded *Siberia* tow five barges from Chicago to Buffalo. The total projected delivery of 256,000 bushels of grain would have set a new record, but it never came together. Rather than a triumph, the *Siberia's* next voyage became a tragedy.

On September 7, 1882, while under the command of Captain Sam Thornton, to whom Davidson had turned over permanent command before the trip, and towing two barges, the *Siberia* encountered a vicious northeast gale while crossing Saginaw Bay. Sometime after midnight, while steering a south-southeasterly course and taking the heavy seas on the beam, a large wave swept the deck and carried second mate C. W. Norris overboard. Norris's absence was not discovered for at least twenty minutes. The time lapse, darkness, and dangerous seas made any rescue efforts impossible. During the worst of the storm, James Davidson bent official rules and took charge of the ship during the crisis. Davidson turned the *Siberia* and its tow into the wind and, using the powerful engine to hold position, jogged into the breaking seas until the safety of morning.[66]

Reaching Buffalo, Captain Davidson wired Norris's widow the following message:

Dear Madame: It is with great sorrow that I write these lines to you. Sept. 7, at 12 o'clock at night, we had a terrible gale of wind about half way across Saginaw Bay, and we shipped tremendous seas, which carried Mr. Norris overboard. I did everything I could to save him, but it was impossible. The sea was so heavy that the barges behind us in tow came very near foundering.

I feel very sorry for losing our beloved friend Norris. I hope the Lord has cared for him. I have advertised in all the papers, and if his body is found I will see that it is taken care of.

Write me, and anything I can do for you, I will be glad to do. When I arrive in Chicago I will call on you and let you know more about the particulars.

Yours truly,
James Davidson
Owner Steamship Siberia. (CIO 13 September 1882)[67]

What happened to the family, or whether Davidson followed up with real assistance, is unknown. Two weeks later, however, Davidson was back in Bay City overseeing the beginning of a new steamer, the *George T. Hope*.

While more research is needed, it is clear that Davidson constantly refined the design of his ships. Having commanded and managed each new vessel as part of his expanding shipping business, Davidson had a breadth of performance data to work with as he contemplated his next build. James Davidson may have had some reservations about the *Siberia*. Despite early reports that his next vessel would be substantially longer than the *Siberia*, the *George T. Hope* was 9 feet shorter and a full 1 foot deeper in draft. Davidson also added a seventh cargo hatch, probably to improve cargo-handling efficiency. Despite the smaller dimensions and gross tonnage (1,559 vs. 1,618 tons), the *George T. Hope* had a cargo capacity equal to, or possibly a bit greater than, the *Siberia*. Both vessels delivered seventy-four thousand-bushel cargoes of wheat during their first seasons of service. That Davidson had specific concerns about the *Siberia*'s design is further borne out by major modifications to increase longitudinal strength made after the 1884 season. These improvements included a new lower deck, bilge braces, and raised bulwarks.[68]

As was his tradition, Davidson took command of the *George T. Hope* during its shakedown voyages, and the ship proved an immediate success. During a forty-day period in September and early October 1883, the *Hope* loaded and delivered four full cargoes, two each of grain and coal, "a feat rarely accomplished with so large a vessel," according to Cleveland's *Marine Record*.[69]

Davidson's next steamer represented a significant jump in hull size and engine power. The *Australasia* had an official length of 282 feet, breadth of 39 feet, depth of 21.2 feet, and a gross tonnage of 1,829. The King Iron Works compound engine was a larger version of those installed on the *Siberia* and *George T. Hope* and, with a reported 2,600 nominal horsepower, was the most powerful engine on the Great Lakes in 1884.[70]

Captain Davidson explained to a visiting newspaper reporter that to achieve the *Australasia*'s great size, along with the usual iron braces, bands, and strapping, he was introducing a new idea in the way of trusses:

One . . . will be placed in the center, about one-fourth way from the stem, another amid-ships and the third about one-fourth way from the stern. The fourth will be placed just forward of the boilers. A heavy truss will begin on either side at the bilge and proceed to the center of the upper deck, and thence to the main keelson. The truss first alluded to will be two and one-half inches in diameter, and will be firmly fastened at top and bottom with nut and washer set up tightly. The braces, from bilge to center, will be secured, screwed by a turnbuckle. The first trusses are each calculated to support a 150-ton strain, and those beginning at the bilge, seventy-five tons.[71]

The *Australasia* was an expensive and thoroughly modern vessel. The engine received its power from vast Otis steel boilers, much larger than those on Davidson's earlier steamers. The vessel was equipped with the latest labor-saving American Ship Windlass Company steam windlass and capstan brought in from Providence, Rhode Island. With Davidson in command during its maiden voyage, the *Australasia* carried a record 3,150 tons of salt into Milwaukee.[72]

Building and commanding four bulk steamers in four years provided Davidson with unparalleled opportunities to personally test new technologies and observe the performance of architectural features and designs of some of the largest wooden vessels of the time. The *Australasia* seems to have been a benchmark in the evolution of the Davidson bulk steamer, and nearly three years passed before he launched another one.

Although building no steamers, the Davidson yard remained very active, launching two three-masted schooners, the 979-ton *Polynesia* and the 928-ton *John Shaw*, in 1885. The *Polynesia* became a consort for the *George T. Hope*, while the latter vessel was sold to a Bay City lumber company.[73] Iron strapped and banded like his steamers, the new vessels had full schooner sailing rigs but were intended for use as schooner barges; they were the first of many such vessels the Davidson yard would produce over the next fifteen years. With a predicted capacity of sixty thousand bushels of wheat or 1.2 million feet of lumber and a cost of $48,000, these craft were designed to generate quick profits. In early November 1885, the *Polynesia* collected nearly $3,000 from one charter of Duluth wheat.[74]

The Great Lakes economy was changing rapidly in the early 1880s, with the Lake Superior iron rush promoting a rapid industrialization of ships and shipping. In 1886, during a period of high demand for ships, Davidson sold his fleet to three groups of men involved in the iron rush. Iron brought a new era and several deadly shipwrecks in Wisconsin waters and elsewhere. The archaeological remains of these wrecks and their stories of maritime industrialization reveal the dark underbelly of Great Lakes life during the closing years of the nineteenth and beginning of the twentieth centuries.

Chapter Nine

SURVIVAL OF THE FITTEST

During the 1880s, industrialization brought rapid changes in the size, composition, and operation of the Great Lakes fleet. In 1880, steam vessels accounted for 38 percent of the tonnage in the commercial fleet. By 1890, that figure was 65.6 percent. The total tonnage of steamers went from 212,045 to 652,923—a 208 percent increase. The number of steam vessels increased from 931 to 1,597, a 71.5 percent increase. By contrast, the official sailing tonnage grew only 9 percent, and the number of sailing vessels in service declined from 1,473 to 1,272, a 13.6 percent drop. Great Lakes shipbuilders continued building commercial sailing vessels, including some of immense size, but the numbers remained modest. Towing the larger sailing vessels was common in 1880 and by 1890 had become the norm. Whether classified as sail or steam, the wooden ships built to carry bulk cargo on the Great Lakes during the 1880s and 1890s became larger, faster, and more efficient.[1]

Beginning in the early 1880s, the boom in Lake Superior iron production created a strong demand for the large wooden schooners built a decade before and stimulated the construction of some new schooners. The typical large sailing vessel built during the 1880s was designed with towing in mind, but most began their service fully rigged and crewed for independent navigation. The fully operational schooner offered ship owners more options in responding to changing market or harbor conditions, even if most of its trips were on a towline.

For sailors, shipbuilders, and businessmen, the Lake Superior iron rush dramatically accelerated the pace of change in Great Lakes marine technology and operations in play since the late 1860s. Iron ore made lucky men rich and rich men even richer. In maritime communities, rapid industrialization brought increased labor unrest and violence and, through the Lake Superior iron trade, the deaths of dozens of sailors.

Iron making had always been important in the United States, and its significance grew throughout the nineteenth century, spurred by the adoption of steam power, the development of railroads, and increasing demands for agricultural and industrial machinery and military equipment. After the discovery of iron deposits in Michigan's Upper Peninsula in the 1840s and 1850s, the Lake Superior region began shipping modest quantities of ore. In 1860, Lake Superior mines supplied 9 percent of the ore used by American pig iron smelters, a figure that grew to 24 percent at the end of the decade. The Civil War and the boom years that followed fueled the demand for iron, and the production of Great Lakes ore grew steadily. In 1873, the Cleveland Iron Company, possibly the largest in operation, shipped 150,000 tons of ore from its specialized docks in Marquette, Michigan. The depression that followed the Panic of 1873 checked but did not stop the expansion of the iron industry.[2]

Among the factors driving the growth in Lake Superior iron mining was the widespread adoption of the Bessemer process for smelting ore by Andrew Carnegie and other large-scale producers during the 1870s. The Bessemer process allowed the production of economically priced, high-quality steel that within a few years replaced puddled cast iron as the dominant structural metal. The Bessemer process also altered the geography of iron mining, as it required the soft, high-phosphorous ores found in vast untapped Gogebic, Vermillion, and Mesabi iron ranges in northern Michigan and Minnesota.[3]

The grain and lumber traffic on the Great Lakes also grew substantially during the 1880s and 1890s, but it was the increasing demand for Lake Superior iron that contributed the most to making these decades especially deadly. In 1880, the Lake Superior region produced 1,677,814 tons of ore, 23.6 percent of the national output. In 1890, the region yielded 10,332,248 tons—over a 600 percent increase during the previous decade and accounted for nearly 65 percent of American ore produced.[4]

Industrialization also resulted in large increases in volume of coal shipped to Lake Michigan and Lake Superior ports. Although shipping rates for coal were typically much lower than for ore during the 1880s, it was important westward cargo for bulk carriers when available, and the twin ports of Duluth, Minnesota, and Superior, Wisconsin, became an important coal distribution hub for the Lake Superior region. For understanding the pressures on Great Lakes ships and sailors, the rate of increase in coal

View from the shoreline toward the three ore boats at the Soo Line Railroad dock being loaded with iron ore. The first in line is the *William B. Pilkey*. WHI IMAGE ID 131297

transportation is more important than the absolute volume carried during the 1880s. The twin ports received 60,000 tons of coal in 1880, 592,000 tons in 1885, and 1,535,000 tons in 1888.[5]

The expanding economy prompted some to build large new steam bulk carriers. By one well-informed estimate, the carrying capacity of the Great Lakes fleet doubled between 1885 and 1890. More often, it was older vessels—typically the large grain schooners and first-generation steam bulk carriers built in the early 1870s that were pressed into service on Lake Superior. It was not only the new ships that increased the fleet's capacity; improved loading and unloading technologies and facilities allowed owners to push existing vessels and crews harder. In the early 1880s, fifteen or sixteen round trips constituted a good season for vessels in the Lake Superior iron trade. By the early 1890s, twenty-two trips were "now considered nothing more than a fair season's work."[6]

One approach to increasing economic productivity in the ore trade during the 1880s was the expanded use of large schooners and schooner barges as "consorts" towed behind steam bulk carriers. On the Great Lakes, the practice began in the 1860s, when firms began using steam tugs and innovative steam barges to tow strings of old schooners with simplified sailing rigs. Some also dismantled obsolete passenger steamers, repurposing them as high-capacity barges, frequently equipped with auxiliary sails. Success with the large barges inspired Cleveland shipbuilder Elihu Peck to design and build the *R. J. Hackett* and its unpowered consort barge, the *Forest City*, in 1869 and 1870.[7] The forty-seven first-generation bulk steamers built between 1869 and 1874 routinely towed one or more large schooners or purpose-built schooner barges.

A great moneymaker, the schooner-barge system earned a dark reputation among America's maritime laborers for low pay, poor working conditions, and excessive risk to life and limb. Conditions for schooner-barge sailors along the US Atlantic Seaboard were worse in general than on the Great Lakes. In both regions, however, the move from independent sailing craft to towed schooner and schooner barge was part of a broader package of rapid industrial transformation that brought great wealth to relatively few maritime capitalists and increased toil, uncertainty, injury, and death to many others.

The Lake Superior iron ore trade of the 1880s was as much of a maritime frontier as western Lake Michigan had been in the 1830s, 1840s, and 1850s. Maritime frontiers attract the dreamers, the bold risk takers, and often the ruthless. On the iron frontier, one such man was Captain James Corrigan (1848–1908), a self-made industrial capitalist who began investing in Lake Superior iron in 1883 and who would, for a time, control one the largest fleets of ore carriers on the Great Lakes.

Like James Davidson, James Corrigan had been forced to grow up early. Born in Ontario in 1848, James and his older brother John left an unhappy home after the death of their mother in 1859. Making their way to the thriving Lake Ontario port of Ogdensburg, New York, the boys began their working lives along the waterfront as Great Lakes sailors. James, the more ambitious and aggressive of the two, acquired the small schooner *Trial*, which he brought to Cleveland, Ohio, in 1867 or 1868.

An important lake port, by the late 1860s Cleveland was also becoming the heart of the American petroleum industry and home to the Corrigans' boyhood friend John D. Rockefeller, who in 1870 organized Standard Oil Corporation. Petroleum was an unregulated wildcat industry, and, perhaps inspired by Rockefeller's rapid rise, James and John Corrigan invested their capital and energy into oil, owning refineries as early as 1872. More than just a sailor or capitalist, James Corrigan proved a talented inventor and developed new refining processes that allowed the firm to produce a variety of specialty oils. By 1875, Corrigan & Company owned and operated the Excelsior Oil Works, one of the largest refineries in Cleveland. The business employed twenty-five men and had estimated annual sales of $300,000. At their peak, James Corrigan and his brother owned at least four refineries, including the largest refining works in the United States outside of the Standard Oil Corporation.

John D. Rockefeller was more of an oil evangelist than a capitalist robber baron. He considered Standard Oil an almost divinely sanctioned entity, created to bring inexpensive light to the world. Seeking to bring order to the wasteful petroleum industry, he used a variety of means to acquire or eliminate competing oil refineries. As muckraking journalist Ida Tarbell revealed years later, Rockefeller used his overwhelming economic leverage over the railroads to extract preferential rates, rebates, and treatment for Standard Oil. Rockefeller forced the railroads into charging higher rates or refusing service to independent refineries such as the Corrigans' Excelsior plant. As did most of the Ohio refiners, the Corrigan brothers capitulated to Rockefeller. In 1879, they leased their plants to Standard Oil and left for Europe, where they spent the next few years developing oil fields and refineries in the Polish region of the Austro-Hungarian Empire. James Corrigan returned to Cleveland in 1883 and sold his refineries to John D. Rockefeller for a large block of stock and a directorship in the Standard Oil Corporation.[8]

James Corrigan then set his sights on Lake Superior iron mining and went into partnership with Franklin Rockefeller, John D.'s personable but troubled younger brother. Corporations and partnerships became common tools in the 1880s, and mapping out James Corrigan's entire web of business relationships associated with mining and shipping and their sequence is beyond the scope of this book. What is significant is that with his forceful personality, James Corrigan seems to have dominated the enterprises he took part in.

James Corrigan kept his hand in maritime commerce even while enjoying his early successes as an oil refiner. During the mid-1870s, Corrigan purchased and operated at least two sailing vessels, the 278-ton bark *Massillon* and the 295-ton schooner *Hippogriff*. The circumstances surrounding Corrigan's purchase of the two ships and the prices he paid are unknown, but their age, condition, and common fate provide evidence of what would become James Corrigan's long-term strategy for extracting money from old ships: buy cheap, run hard, and insure heavily.

Steely-eyed Cleveland capitalist James Corrigan (1848–1908) made a fortune on the Great Lakes, but at high cost in sailors' lives. CLEVELAND PUBLIC LIBRARY PHOTOGRAPH COLLECTION

Built in 1856, the *Massillon* was an old and hard-used schooner that had survived a long series of groundings, collisions, and near foundering in the years leading up to Corrigan's purchase in the late spring or early summer of 1876. When Corrigan or his agent contracted to carry a load of coal to Lake Michigan, the *Massillon* had been inactive for a least a month. Left unattended and not well maintained, the hull above the waterline dried out, causing the seams between the exposed planks to open up. Encountering rough weather after leaving Cleveland, the *Massillon* began taking on water through the plank seams at an alarming rate. With hard work at the bilge pumps and some luck, the vessel made it to Detroit in near-sinking condition.[9] After some minor recaulking, the vessel continued in service. Just four months later, bound for Chicago carrying a heavy load of limestone, the *Massillon* sprang a leak in good weather and sank in Saginaw Bay. The crew was lucky. Although the incident occurred at night, a passing tug spotted their distress signal and rescued them. James Corrigan had insured the vessel and cargo and apparently recovered most or all of his original investment on top of the profits made over the summer.[10]

The *Hippogriff* was a canal schooner built in Buffalo in 1863 to carry grain. A newer but by no means new vessel, under Corrigan's ownership the schooner continued to carry grain and coal but also lumber, limestone, and pig iron. On August 25, 1877, the *Hippogriff* lost its jibboom, bowsprit, head gear, and two headsails in a collision with the tug *Prince Albert*.[11] A month later, the vessel limped into Door County's North Bay with a major leak.[12] After minor repairs, the *Hippogriff* continued to Chicago and took on a large cargo of oats. Sailing up the lake about twenty miles from Kenosha, Wisconsin, it collided with the schooner *Emma L. Coyne* and sank. Once again, the crew proved lucky and escaped, this time onto the deck of the *Coyne*. James Corrigan carried $6,000 of insurance on the schooner and $8,000 on the cargo and suffered little if any real loss.[13]

Managing the *Massillon* and *Hippogriff*, James Corrigan deliberately or unconsciously employed one of the dark strategies of maritime capitalism—the squeezing of voyages out of old or poorly maintained ships. Shipwreck anthropologists have coined the "one more voyage" thesis to describe the practice of owners continuing to use vessels until they sank or simply could no longer function.[14] Simply put, without intervening factors, the owners of an old and worn-out ship could charge the same rates for hauling cargo as a more expensive and appropriately maintained first-class vessel. This practice was especially common in shipping trades where the cargo was robust and of low per-unit value, such as coal, ore, and timber products.

Under these conditions, the crew, not the owner, assumed the real risks to life and limb that no insurance policy could mitigate.

When James Corrigan began to buy ships in anticipation of his iron mining business, he again invested in older, less expensive craft. In 1884, he purchased the schooner *Niagara* from the H. J. Winslow estate for $31,000.[15] At nearly 205 feet in length and 766 gross tons, the three-masted schooner was among the largest of its class when launched by Parsons & Humble at Tonawanda, New York, in 1873. During his first two years of ownership, Corrigan largely employed the *Niagara* in the same trades as its original owners: mostly hauling grain and coal—sometimes to Lake Superior but usually only as far west as Chicago or Milwaukee. Typically, the schooner was towed as a consort by large new steam bulk carriers such as the *City of Rome*, but it also made partial and entire voyages under independent sail.

Ore traffic on Lake Superior was moribund when Corrigan purchased the *Niagara* in 1884, with freight rates near their lowest point since the trade began in the 1850s. The rates remained relatively low in 1885, with ship owners receiving between $1.07 and $1.65 per ton for carrying iron ore from Ashland, Wisconsin, to the Lake Erie ports.[16] With the national economy on the upswing and with new mines coming online, however, Corrigan was betting on the future. With predictions for the coming 1886 season strong, he doubled down on his bet, paying $28,000 for the schooner *James Couch*. At 215 feet long and 979 tons, the *James Couch* was one of the largest schooners on the lakes when launched at Marine City, Michigan, in 1871. In an effort to maximize cargo capacity, the vessel had originally been equipped with two leeboards attached to its sides rather than a centerboard. The innovation proved unsuccessful, leading a traditional centerboard to be installed. However, the additional stresses on the hull anticipated in designing for the leeboards led to the construction of a schooner of exceptional strength. Corrigan and his partners also paid $20,500 for the 727-ton *Lucerne*, another large Parsons & Humble schooner nearly identical to the *Niagara*, and $40,000 for the 980-ton *Raleigh*, an early *R. J. Hackett*–style steamer built in Cleveland by Quayle & Martin in 1871.

From Corrigan's perspective, buying these ships was good business. He had made money buying older vessels in the past and, at least in this instance, the ships had been strongly built and well maintained. The purchase, however, would ultimately lead to multiple tragedies as the schooners *Lucerne*, *Niagara*, and *James Couch* would each wreck, killing their entire crews while carrying iron ore for James Corrigan on Lake Superior.

Built in 1871 and purchased by James Corrigan and partners in 1886, the *Raleigh* was a first-generation steam bulk freighter designed to carry cargo and tow large vessels such as the schooner *Lucerne*.
C. PATRICK LABADIE COLLECTION / THUNDER BAY NATIONAL MARINE SANCTUARY, ALPENA, MI

The business economics were simple, if sinister. For an outlay of less than $120,000, James Corrigan entered the 1886 season with four vessels and a combined cargo capacity of over 5,000 tons of iron ore. By comparison, Captain James Davidson had spent roughly the same amount of money building his "marine monster" *Australasia* in 1884. Corrigan's 1886 fleet had more than twice the cargo capacity of the *Australasia* and offered more flexibility to respond to fluctuating market conditions, accidents, or unexpected delays loading or unloading cargo—all common events on the Great Lakes. While an idle $110,000 bulk freighter could fast become an economic catastrophe, one or two inactive $20,000 to $30,000 schooners represented just slow business if the rest of the owner's floating capital remained working.

The promising 1886 season began a bit slowly but improved over the summer. On most but not all trips, the *Raleigh*, *Niagara*, and *Lucerne* traveled together under the charge of the *Raleigh*'s captain, William Mack, to whom Corrigan had sold a one-ninth interest in the three vessels at the beginning of the season. In mid-April, an independently sailing *Lucerne* grossed a little more than $2,000 carrying 49,500 bushels of wheat from Chicago to Buffalo. In mid-May, the three vessels carried coal from Buffalo to Chicago for sixty cents per ton—a relatively short trip, with gross receipts of about $2,300. In mid-June, the three ships entered the Lake Superior iron ore trade and made several trips from Marquette, Michigan.[17]

The ore freight rates from Marquette began that season at about $1.10 to $1.15 per ton, and the combined freight receipts for the three vessels was about $4,000 per trip.[18] The rates rose slightly to $1.25 per ton in July. Running light (with no cargo) westbound to Lake Superior was relatively common for ore vessels. However, Corrigan's fleet may have been delivering some coal to Marquette or perhaps to the Duluth Lime and Coal Company, a new venture started by James's brother John Corrigan in April and in which James Corrigan was also a principal stockholder. Coal freights for Lake Superior in midsummer hovered around fifty cents per ton—enough to offset all of the typical operating expenses of a round trip from Ohio, making the ore cargo deliveries essentially all profit.

As the season progressed, a growing demand for ships in both the grain and ore trades began to drive up freight rates. By late August, the rate for ore from Marquette to Cleveland had risen to $1.50 per ton. On September 5, it reached $1.60, and the next day an ore carrier was chartered for $1.75 per ton. On September 13, the Marquette rate climbed to $1.85, while the rate from Ashland, Wisconsin, rose to $2.35 per ton.[19] The August 30, 1886, issue of the *Chicago Inter Ocean* proclaimed that "the present boom in the marine business has proven conclusively that there is not enough tonnage on the lakes. From every shipping point comes a demand for more tonnage. Agents are harassed on every hand with telegrams and letters, and they have almost exhausted themselves and their fleets, too."[20]

James Corrigan had indeed made an excellent bet at the beginning of 1886 when he bought the three older vessels. As of September 26, 1886, Lake Superior had shipped 30 percent more iron ore than at the same date in 1885, and freight rates on coal, lumber, and ore were higher than at any time in the previous five years.[21] This translated into fast money for Corrigan and his subordinate partners.

On October 9, the *Raleigh*, *Niagara*, and *Lucerne* landed coal at Duluth for $1 per ton. On October 13, they cleared Ashland, Wisconsin, for Cleveland loaded with ore at an estimated $2.60 per ton. Assuming a combined load of at least 3,500 tons, the round trip from Cleveland, Ohio, to Lake Superior generated at least $12,600 in gross receipts, or 14 percent of the price Corrigan had paid for the three vessels just ten months before. Such high profits encouraged high risks. During this most profitable leg, the *Niagara* and *Lucerne* were battered by heavy weather and arrived at the Sault

Locks in "a leaking condition."[22] After making quick repairs they sailed on without incident to Cleveland.

On the next trip, Corrigan's flotilla once again carried coal to Lake Superior, with the *Lucerne* splitting off to deliver its cargo in Washburn, Wisconsin, before joining up with the other vessels to load iron ore at Ashland. By the time the *Lucerne* reached Ashland on November 12, shippers were offering charters of $3 per ton and possibly more. On November 15, the *Lucerne*, under the command of Captain George Lloyd, was loaded with 1,256 tons of ore. The *Lucerne* had taken a beating on the previous trip, and perhaps Captain Lloyd or Corrigan's vessel agent was growing cautious, for this was more than 100 tons less than the schooner had typically carried on earlier voyages during the season. Even with the lighter load, the *Lucerne*'s round trip carrying coal from and iron ore to Cleveland stood to gross Corrigan and Captain Mack at least $5,000, a nearly 25 percent return on the $20,500 they had paid for the schooner.

The weather was excellent on the evening of November 15, 1886. With the barometer readings high and holding steady, Captain George Lloyd elected to strike out alone and under sail for the Sault Locks, where he would meet up with the *Raleigh* and *Niagara*. With freight rates at stratospheric levels, and the end of the navigation season looming, the decision to get under way quickly made sense. Lloyd's personality also may have factored into the decision. According to an account appearing in Cleveland's *Marine Record* two weeks after the tragedy, Lloyd had tremendous pride in his schooner and "declared he would rather go out with her under canvas than under the tow of any steamer."[23]

Threading the needle up from Ashland to deep water would require a lot of sail handling and was probably not a popular decision among the eight crewmen Lloyd commanded, and sailing conditions must have been excellent. The port of Ashland, Wisconsin, is located at the southern end of Chequamegon Bay; the bay and port are well protected by a seven-mile-long spit of land ending at Long Point. Entrance to the bay is through a narrow channel that separates Long Point and Madeline Island, the most southerly and the largest of the Apostle Islands. Outbound, the most direct route from Ashland to the Sault Locks involved sailing for ten miles on a course of about eighteen degrees true north and then due west following the "South Channel" for about three miles, passing between the southwest tip of Madeline Island and the north side of Long Point. The deep-water channel between these two headlands narrows to less than a mile. Extensive sand shoals on both sides can make passage hazardous, especially in limited visibility. The outbound leg proved uneventful, with the *Lucerne* safely reaching open water.

The storm of November 16 and 17, 1886, struck without warning and became one of the most vicious in Great Lakes history. Not only had mariners missed its signs, the daily maps and forecasts of the US Army Signal Corps (precursor to the US Weather Bureau) failed to catch a large low-pressure system and cold front developing southwest of the Great Lakes.[24] The November 15 map, based on 7 a.m. observations, depicts a stable area of high pressure across southern Lake Superior, with light westerly winds and temperatures just below freezing. The map issued on the morning of November 16 continued to report stable good weather. At Duluth, Minnesota, the temperature was twenty-nine degrees Fahrenheit, with light winds from the west and a barometric pressure of 30.47 inches. Slightly more than two hundred miles to the east, Marquette, Michigan, reported thirty degrees Fahrenheit, west winds at fifteen miles per hour, and a barometric pressure of 30.31 inches. On paper, the weather looked nearly perfect for the *Lucerne*'s voyage.

The weather map for the morning of November 18 presented a radically different picture. The temperature at Duluth had plummeted to sixteen degrees and the barometric pressure to 29.48 inches. To the east, neither Marquette nor Escanaba, Michigan, reported in that morning, the storm having probably knocked out the telegraph lines. The rapid changes in temperature reported south of Lake Superior during the previous twenty-four hours provide an even greater sense of the power associated with this storm and its rapid onset. The temperature at Chicago had fallen by twenty degrees, Milwaukee by sixteen degrees, and Davenport, Iowa, by twenty-four degrees.

For Captain George Lloyd and the *Lucerne*, conditions soon went from excellent to troubling and then became a nightmare. In the roughly twenty hours after leaving Ashland, fair and friendly winds carried them some sixty to seventy miles to the west of the entrance to Chequamegon Bay. In midafternoon, the wind abruptly shifted to the northeast, making further westerly progress impossible. At four o'clock in the afternoon on November 16, Captain Bassett of the steamer *Fred Kelly* spied the vessel off of Ontonagon, Michigan, with all sails except the fore-gaff topsail set. Bassett described the *Lucerne* sails as all "hanging loose," suggesting that the schooner was "in irons," a dangerous position where the vessel is unable to steer and thus especially vulnerable in the heavy seas. As darkness closed in, the mate of the *Fred Kelly* watched the *Lucerne* finally come about and head downwind back toward the shelter of Chequamegon Bay.[25]

Soon the *Lucerne*'s crew was fighting twenty-five- to thirty-foot-high seas generated by sixty- to seventy-mile-per-hour winds passing over two hundred miles of open lake. Taking waves on the starboard quarter and directly on the stern, the *Lucerne* would have been difficult to steer and in ever-present danger of being pooped by one of the giant waves smashing down onto the quarter deck, or of being broached if the bow fell too deeply into the trough between the steepening seas.

A well-founded and competently handled sailing vessel can survive extraordinary storms when far offshore. Captain Lloyd had more than twenty years of sailing experience and, by most accounts, the *Lucerne* was in good condition, with new decks and hatch coamings in 1885 and new sails at the beginning of the 1886 season. What Captain Lloyd lacked was sea room. The closer the *Lucerne* came to sheltered water, the higher the danger of the schooner being driven uncontrollably onto what sailors call a *lee shore*. Short of a fire, it is probably the worst nightmare for the crew of a storm-tossed sailing vessel.

What specifically prompted Captain Lloyd to turn the *Lucerne* about that evening will never be known, but the situation must have been desperate. Perhaps he believed that the sudden storm would quickly blow itself out. Maybe the battered *Lucerne* was taking on water. Possibly Lloyd simply took it for granted that he possessed the skill required to make the dangerous passage back into Chequamegon Bay. Whether he tried to check his schooner's westward progress toward the potentially deadly shore by tacking his ship in a series of zigzag courses is unknown. It may be that the waves, ice, or other circumstances made the sail handling required for such course changes too dangerous.

If Captain Lloyd kept to the course observed by the *Fred Kelly*, reaching the entrance to Chequamegon Bay would have taken, at the most, ten or twelve hours, probably less. This placed Lloyd and his crew in a harrowing situation of making landfall in the dark during a blinding snowstorm. The closer the *Lucerne* came to shelter, the fewer options Captain Lloyd had for altering his course. Seven miles east of the south passage between Long Island and Madeline Island, the *Lucerne* had about ten miles of deep water in which to maneuver; at three miles out, the *Lucerne* had less than five. Once in the south channel, the *Lucerne* had to keep within a deep-water corridor less than a mile wide.

Approaching the south channel in blinding conditions, Captain Lloyd would have been dependent almost exclusively on the lead line to ascertain his position. His wet and freezing men would have been casting the lead on

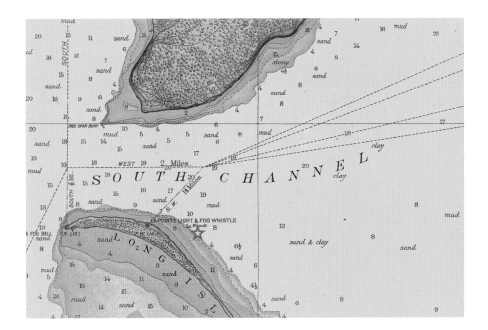

This detail from a 1906 nautical chart of the Apostle Islands shows the narrow South Channel into Chequamegon Bay. The red star marks the wreck site of the schooner *Lucerne*. Note the proximity of the location to deeper water. NOAA OFFICE OF COAST SURVEY

a regular, and eventually on a frantic, basis. If they could keep the *Lucerne* sailing over a muddy bottom in water deeper than fifteen fathoms (ninety feet), they would pass safely into sheltered water. Roughly three miles from the entrance to the Bay, Lloyd's luck failed. The water began growing shallow and the bottom increasingly sandy. Probably when the lead reported a depth of about ten fathoms (sixty feet), Captain Lloyd ordered the ship to come about into the wind and seas and dropped the heavy anchor.

The banging of the heavy links passing out through the chain locker, mixed with the noise of the wind and slapping rigging, created a cacophony that could have rattled the nerves of the most hardened sailor. Although Lloyd and his crew could not see the deadly shore, the shape of the waves and their soundings told them that it was too close. The men at the windlass watched the anchor chain play out, and as the schooner slipped backward, its bow pitched up and down straight into oncoming waves. At the captain's signal, they engaged the pawls on the windlass to stop the outward play of the chain. As the weight of the *Lucerne* came to bear, the chain tightened and the anchor bit into the bottom, perhaps jumping once, twice, or three times before taking hold. The *Lucerne* carried 80 fathoms (480 feet) of chain, enough for a long scope, which would increase the chances of holding and dampen the ship's up-and-down motions. Bringing the *Lucerne* successfully to anchor was no small accomplishment, and Lloyd had made it far enough west into the Apostle Islands to gain some shelter from the storm. However, once anchored, Lloyd and his crew were out of cards. All they could do was keep the bilge dry, stay warm, and pray that the anchor held and that their ship could endure the continued strain.

On Thursday, November 19, after the *Lucerne* failed to appear at the Sault Locks, Captain Mack of the steamer *Raleigh* telegraphed John Corrigan in Duluth asking him to dispatch a search party. On Friday, November 20, the tugs *Cyclone* and *S. B. Barker* left Bayfield, Wisconsin, to look for the schooner. The weather remained foul, and wind, waves, and snow made the going difficult; however, they did not have to travel far. Little more than five miles from Bayfield, just off shore of Long Island, they spied the *Lucerne*'s three tall spars projecting out of the water. Coming closer, they found what the *Marine Record* reported as "literally a ship of death," with three ice-encrusted sailors hanging in the rigging.[26]

Identifying the nine or ten men who died on the *Lucerne* took weeks, and the results were uncertain. At least four sailors, all believed to be young men, were never found. In an effort to identify the victims, the *Ashland Weekly Press* published a detailed description of the three frozen sailors. Two were estimated to be in their early forties, one with a heavy beard and the other clean-shaven except for a mustache. Both wore rubber boots and were heavily clothed. The mustachioed sailor seemed to be wearing every stitch of clothing he owned, five overcoats layered over heavy underwear. The third body, that of a slightly built young man, wore boots but was lightly dressed. The body of mate Robert Jeffreys washed ashore a bit later. The following summer another body, believed to be from the *Lucerne*, came ashore on Long Island.[27] When informed of the *Lucerne*'s loss, James Corrigan issued an order to have the dead sailors "decently interred" at the company's expense. Burying four men was a small business expense on the Lake Superior iron frontier. To be fair to Corrigan, it was almost universally described as one of the worst storms that had ever been experienced on Lake Superior, and it claimed as many as forty ships and fifty lives across the Great Lakes.[28]

James Corrigan had a banner year in 1886 with his $120,000 fleet of four well-used wooden ships generating tens of thousands of dollars in freight revenue. Although men died, the sting of the *Lucerne*'s loss was partially removed by a $20,000 marine insurance policy. Corrigan's financial return on the ships was even greater if one includes the value of services they provided to his iron mining and coal/limestone businesses. In a season when the price of iron ore was high and the ships to carry it scarce, Corrigan could always get his product to market.

Archaeological Evidence on the *Lucerne*

Today the wreck of the *Lucerne* rests in shallow water inside the boundaries of the Apostle Islands National Lakeshore. During the 1980s, a team of divers working under the supervision of professional archaeologists recovered more than four hundred artifacts and completed a preliminary documentation of the wreck. The composition of the artifacts and their pattern of distribution across the wreck underscore the extreme social divide characteristic of commercial sail—a divide widened by the process of maritime industrialization fully underway when the *Lucerne* wrecked in 1886. Among the artifacts recovered from the stern of the vessel—which housed living quarters for the captain, mates, and cook and a dedicated stateroom for owners or important guests—were high quality crockery for dining, porcelain doorknobs for the cabins, a fine pair of binoculars, and an expensive gilt-covered Bible embossed with the name *Lucerne*. Such material remains represent a relatively affluent and well-ordered middle-class life. The artifacts recovered from the forecastle—the dark, crowded, and perpetually damp space in the extreme bow of the ship where the common sailors slept and ate—reveal a rougher working-class lifestyle. Amongst the rough textile cloths and crude eating implements were a rat trap, a liquor bottle, and a crude syringe for injecting opiates.

A series of professional underwater archaeological studies carried out by the Wisconsin Historical Society and faculty and graduate students from East Carolina University documented the hull architecture and surviving machinery. In combination with the previously recovered artifacts and historical research, the 1990 documentation of the *Lucerne* revealed important details about those days between the final sighting of the *Lucerne* by the crew of the *Fred Kelly* on the evening of November 16, 1886, and the discovery of the wreckage on the morning of November 20. The exact position of the wreck relative to the shore offered an important clue. The *Lucerne* is located close to shore in an area of rapidly shoaling water. The water at the wreck is about 24 feet deep and is much too shallow to deliberately anchor a large schooner battling wind and heavy seas. Moving offshore from the wreck, the water deepens quickly to about sixty feet and then to more than one hundred feet in the South Channel. Blinded by darkness and weather and alert to the shallowing of the water, Captain Lloyd dropped the *Lucerne*'s anchor in a spot that he believed was farther offshore than the later events revealed.

Archaeological site plan of the *Lucerne*

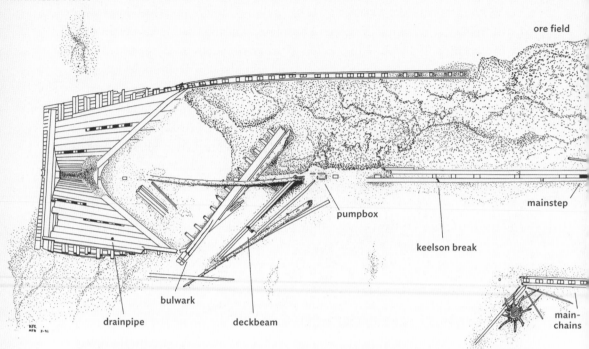

ore field

pumpbox

mainstep

keelson break

bulwark

drainpipe

deckbeam

main-
chains

A. Frozen in time, the crowbar used unsuccessfully in the final moments by the crew to try to save the *Lucerne* and themselves protrudes from the anchor windlass. TAMARA THOMSEN, WHI IMAGE ID 140128

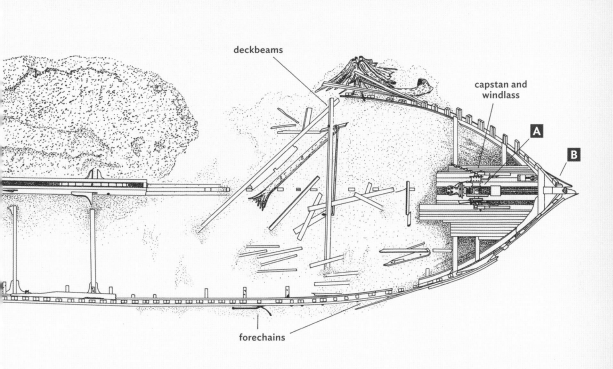

deckbeams

capstan and
windlass

A

B

forechains

B. The bow, stem post, and house pipes of the *Lucerne*. TAMARA THOMSEN, WHI IMAGE ID 140127

Recovered from the forecastle in the 1980s, a stove whose partially burned contents included bits of paneling stripped from inside the ship provided clues about the general quality of life on the *Lucerne* and the final circumstances surrounding her destruction. The bits of unburned ship's paneling found inside the forecastle stove suggest that the *Lucerne* labored at anchor for quite some time before wrecking. The *Lucerne* had a reputation for skimping on groceries and other sailor-related expenses and had not provided the crew with adequate fuel to keep the forecastle warm for what proved an unexpectedly long and cold voyage. This forced the crew to scavenge bits of paneling to keep warm. That the end came quickly is apparent from the contents of the stove and the discovery of only three bodies in the rigging. The heavy clothing on two of the bodies suggests that these men were on deck acting as an anchor watch.

A third clue comes from the powerful windlass, a heavy device located on the bow used to lift and secure the anchor. Archaeologists documented a heavy crow bar twisted up into the windlass machinery. This could only be the result of a desperate and unsuccessful final effort to prevent additional anchor chain from paying out. The *Lucerne*'s hull provides additional information. The keelson assembly (the ship's backbone) is fractured, and the centerboard is fully deployed and embedded into the lake bottom. Great Lakes shipbuilders expected their vessel to repeatedly go aground and designed the centerboard assembly to push back up into the centerboard trunk if it encountered lake bottom while under sail. What they did not expect is to have a centerboard in the deployed position when a ship was moving backward through water. The position of the centerboard on the wreckage indicates that the *Lucerne* was traveling backward when it struck bottom.

The bits of archaeological and historical evidence add up to tell a story. Sailing blind and encountering shallowing water while approaching land, the captain realized that he could not make safe passage into Chequamegon Bay. Turning the ship when the water depth was less than one hundred feet, he ordered the dropping of the anchor in what turned out to be perhaps fifty feet of water. Not realizing how close he was to shore, he kept the centerboard deployed to help steady the *Lucerne* as it rode at anchor in rough water and high winds. After many hours and more likely a day or more, the stress of the surging waves and press of wind on the load schooner caused a sudden catastrophic failure of the windlass. As the two men on watch worked to jam the crowbar into the windlass and check the paying out of the anchor chain, the *Lucerne* was sliding backward toward shallow water. Riding up a large wave, the *Lucerne* slammed into the lake bottom, landing on top of the deployed centerboard. This trauma broke the ship's back and sent the *Lucerne* immediately to the bottom, trapping all but three of the crew below. The two men on watch and another sailor who, dressed in rubber boots may have been preparing to go deck, avoided drowning by springing into the rigging but soon froze to death, covered by inches of ice.[29]

MORE IRON, MORE SHIPS, AND MORE TRAGEDY

The overwhelming success James Corrigan was enjoying in 1886 prompted him to expand his fleet, this time adding first-class vessels. In late September, Corrigan and his new partner, John Huntington, paid James Davidson $160,000 cash for the *Australasia* and the 940-ton iron-strapped schooner *Polynesia*. These were expensive modern ships and suggest Corrigan's growing confidence in the Lake Superior iron trade. Corrigan apparently liked the ships and doing business with Captain Davidson. A few weeks later, Corrigan and Huntington paid Davidson $10,000 for a recently repaired 1873-built scow schooner, the *R. J. Carney*, then committed to buying the two large bulk steamers under construction at the Davidson shipyard.[30]

With the *Australasia*, Corrigan and Huntington went into 1887 as owners of one of the finest wooden bulk steamers on the Great Lakes. The three-year-old vessel was not the largest bulk steamer on the lakes, but its twenty-six-hundred-horsepower compound steam engine made it almost certainly the most powerful in the fleet. Davidson had installed the best towing and lifting equipment available, including self-lubricating American Ship Windlass Company steam windlass and capstans on the bow and on the stern. The addition of a stern windlass was a Davidson innovation, reportedly "capable of doing the work of twenty men in heaving around the docks." The additional windlass also helped in handling the tows of large schooners and schooner barges that were integral to the steamer's operations.[31]

With the Corrigan schooner *James Couch* in tow, the *Australasia* completed the 1887 season's first round trip between Buffalo and Chicago. Winter retained something of a grip, however. The *Australasia* and *James Couch* changed course to avoid ice and ran aground about six miles south of Mackinac City. The tug *John Owen* released the undamaged *Australasia* at a cost of $500. Profits seem to have justified the risk and expense of the tug, as the voyage prompted a reporter in Cleveland's *Marine Record* to congratulate the *Australasia*'s Captain Cowen for "the business qualifications he has shown for his great charge."[32] The following trip, the *Australasia* loaded seventy-seven thousand bushels of wheat at Green Bay, a local record. At 4.25 cents per bushel, the trip back to Buffalo generated a little over $3,000 in freight receipts. On the following trip, the 800-ton schooner *Minnehaha* veered off its tow and almost ran head-on into the *Australasia*, destroying about eighty-five feet of bulwarks and rail and damaging a few planks. After stopping to pick up repair materials, the staunch steamer and its consort continued on to Duluth. The offending *Minnehaha*, however, required major repairs.[33]

Throughout July and August 1887, the *Australasia* made multiple round trips between southern Lake Superior and Lake Erie, usually towing the schooners *Niagara* and the *James Couch*. The freight rates remained stable and high, fluctuating from $2.25 and $2.45 per ton from Ashland to Cleveland. The westbound rate for coal into Duluth was typically seventy-five cents per ton. The *Australasia* could carry more than 2,200 tons of ore or coal, the *Niagara* and *James Couch* each about 1,400 tons, giving the three a combined capacity of roughly 5,000 tons. At the prevailing rates, the *Australasia* and its consorts grossed nearly $16,000, sometimes more, for each round trip—about 10 percent of the original price Corrigan and his partners paid for the three ships. Corrigan was doing equally well on the Lake Superior run with the older steamer *Raleigh* and the practically new 980-ton Davidson-built schooner *Polynesia*, the replacement for the wrecked *Lucerne*. Compared with the bulk steamer *Raleigh*, which had towed the *Niagara* successfully for three seasons, the *Australasia* was a massive craft, with double the gross tonnage and more than twice the engine power. The steamer was heavier and could pull harder and potentially place more strain on the timbers of the older schooners than the smaller *Raleigh*.

On Monday, September 5, 1887, the *Australasia* and *Niagara* left the docks at Ashland, Wisconsin, loaded with iron ore. Commanding the *Niagara* and its nine crewmen was Captain Henry Clements, a veteran mariner from Cleveland who had sailed the Great Lakes for most of his working life. He left the port in a state of disappointment, the plan to have his wife and five young children sail with him having fallen through at the last minute. The *Niagara* was carrying a heavy cargo, perhaps too heavy. While Cleveland's *Marine Record* (published in Corrigan's hometown) reported the weight as 1,400 tons, the *Chicago Daily Tribune* printed a figure of 1,545 tons. Whatever the actual figure, the consensus in surviving sources is that the *Niagara* carried an unusually heavy load.[34]

September 5, 1887, was a grayish day in Ashland, with light rain and a comfortable morning temperature in the low sixties Fahrenheit. The barometric pressure was above thirty inches but had been slowly falling over the past twenty-four hours, and the daily forecast issued by the US Army Signal Office predicted "cooler, threatening weather and local rains; (with) light to fresh winds, generally southerly."[35] Not a perfect forecast, but not a severe one, either. The *Australasia* and *Niagara* left sometime during the day, and like the schooner *Lucerne* the year before, they set their course for the canal at Sault Ste. Marie, the gateway to Lake Superior more than three hundred miles in the distance.

The *Australasia* and *Niagara* made slow but steady progress, steering a northeast by easterly course for more than 160 miles around the tip of the Keweenaw Peninsula. The weather gradually deteriorated throughout the journey, but the tow successfully weathered Keweenaw Point and changed to an east-southeast course to pass Whitefish Point, roughly 135 miles away. By the time they cleared the Keweenaw, the two ships were battling a full gale, with strengthening northwest winds and rapidly building seas. For more than 120 miles, the ships labored hard, with the wind and high seas coming from almost dead astern, making the steering difficult and overall conditions more dangerous.

On Wednesday morning, September 7, disaster struck when the towline between the *Australasia* and *Niagara* parted. Accounts of the wreck printed in various newspapers varied widely. Some reported that in the growing distress, Captain Crowley of the *Australasia* deliberately cast loose the schooner. An account attributed to Captain Crowley printed in the *Chicago Inter Ocean* stated, "*Niagara* suddenly collapsed and sank, seemingly without warning. Whether the strain of the bow line tore out her bow, or her back was broken while her ends were simultaneously elevated by the seas, or whether she was simply swamped by the rush of the waves is not known and probably never will be."[36] Thirty-five years later, the *Buffalo Courier* printed the recollections of "Sailor Sam" Swanson, an old mariner visiting Buffalo who shipped aboard the *Australasia* on the tragic voyage. According to Swanson,

> Both boats labored heavily, and when about 10 miles above Whitefish the towline parted. The schooner had no canvas to steady her, and she fell off into the trough of the sea. I saw her lying over on her starboard side for a moment with her lee rail under water, and waves sweeping over her. Suddenly one immense sea came along and she was lifted on top of it. As she righted, there was an awful crash, and her spars went toppling down, tearing up the deck and crushing the bulwarks as she fell into the terrible sea that was running.[37]

Owned by James Corrigan and partners, the James Davidson-built *Australasia* had one of the most powerful steam engines on the Great Lakes. The iron-laden Corrigan schooner *Niagara* wrecked with all hands when the tow cable to the *Australasia* parted in a violent storm on September 1887. C. PATRICK LABADIE COLLECTION / THUNDER BAY NATIONAL MARINE SANCTUARY, ALPENA, MI

A contemporary account printed in Cleveland's *Marine Record* told the same story but added that a gust of wind blew out the foresail and mainsail, making it impossible for Captain Clements to keep his ship pointed into the immense seas. Clements and his men worked furiously with axes to clear the fallen masts from the schooner. Realizing that the ship was sinking fast, the crew made a desperate effort to escape in the yawl but were almost immediately consumed by the waves. There were no survivors.[38]

When the *Lucerne* went down the year before, virtually no one blamed the captain or the owner. That was simply a shipwreck with brave sailors surprised and overcome by a deadly storm. The September 6–7, 1887, storm that claimed the *Niagara* was not quite as powerful as the November 1886 storm, but some mariners described it as the worst they had ever encountered. The September 1887 storm, however, was not without some warning, and the *Detroit Free Press* concluded, "It is certain that the boat was overloaded, and her fate should be upheld as an example to vessel captains, no matter how anxious they be to please owners."[39]

Captain Clements may indeed have been desperate to please. He apparently had no savings, and his death left his family, which included his wife, invalid mother, and five children ranging in age from one to ten years of age, destitute. There was little safety net for the working mariner but plenty for the bold capitalist. James Corrigan carried $25,000 in insurance on the *Niagara*, only $6,000 less than he had paid for the ship in 1884. In less than three years, against a loss of $6,000 and ten men's lives, the *Niagara* had earned Corrigan and his partners tens of thousands of dollars. Five weeks after the disaster, the *Marine Record* ran a brief note praising James Corrigan's beneficence:

> The generosity of the owners of the lost schooner *Niagara* is worthy of emulation in these days of disaster on the lakes. About a week ago Messrs. James Corrigan and John Huntington headed a subscription paper for $200 each for the widow of Captain Clement and Mr. John Corrigan added $100.00. The list now bears a total of $800. This and what has been taken up by J. P. Deveny will swell the sum to about $1300. Mrs. Clements was left destitute with five small children.[40]

In supreme irony, this bit of news was printed just below a report of the finding and burial of the body of George Lloyd, the long-missing captain of the *Lucerne*.

Corrigan invested heavily in ships in 1887. In June, he paid $75,000 for the five-masted 1,418-gross-ton schooner *David Dows* and $45,000 for the four-masted 1,443-gross-ton schooner *George W. Adams*, both vessels built at Toledo by the Bailey brothers, James Davidson's former employers.[41] These massive schooners had some of the largest carrying capacities of any lake vessel afloat. Corrigan and Huntington also purchased two new bulk steamers under construction at the Davidson yard. They took delivery of the 1,837-gross-ton *Roumania* in May and the 1,888-gross-ton *Bulgaria* in September, paying $140,000 for each steamer. They also took delivery of the 2,282-gross-ton bulk steamer *Aurora*, at the time the largest and one of the most powerful wooden bulk carriers on the Great Lakes.

The cordial business relationship between Corrigan and Davidson came to an abrupt end with the loss of Corrigan's almost new Davidson-built schooner *Polynesia*, which sank in Lake Michigan, about twenty miles east of Sheboygan, Wisconsin, on October 24, 1887. Conditions were terrible, with wind blowing out the *Polynesia*'s sails and the brutal waves rolling the mainmast out of the consort steamer *Raleigh*, nearly killing some of the crew. Without sails or its steam consort, the heavily laden *Polynesia* was helpless. Laboring heavily and shipping tremendous volumes of water over the bow, the vessel slowly began to sink. With the *Polynesia*'s bow and lee rail underwater, Captain J. W. Kerry ordered his crew to launch the yawl and abandon ship. Minutes later, Captain Read of the *Raleigh* cut the towline and watched the *Polynesia* go down. Read managed to rescue the *Polynesia*'s crew, but it was a very close call.[42]

Although the storm claimed multiple vessels and several lives on Lakes Huron and Michigan, the loss of the *Polynesia* came as a shock. Compared with the *Lucerne* and the *Niagara*, the *Polynesia* was practically new, and with all the advanced architectural features of Davidson vessels, including iron arches and straps, it was an extremely strong vessel. The *Detroit Free Press* placed the blame squarely on Davidson, stating that the *Polynesia* "was a victim of that folly which many of the leading ship architects persist in—that is, her bulwarks were built as solid as a stone wall," which did not allow the vessel to shed the water crashing over the bow with every heavy wave.[43] Solid bulwarks *may* have been a contributing factor, but given the ship's slow settling, it seems likely that a plank or seam gave way along the waterline or below the surface of the hull. The solid bulwarks contributed to great longitudinal strength required on the larger Davidson vessels.

Three months after the loss of the *Polynesia*, an attack on Davidson appeared on the first page of the January 19, 1888, issue of Cleveland's *Marine Record* in the form of an unsigned account of repair work under way

Sold to James Corrigan and partners by James Davidson in 1887, the new *Roumania* became the first Great Lakes freighter powered by a triple-expansion steam engine. C. PATRICK LABADIE COLLECTION / THUNDER BAY NATIONAL MARINE SANCTUARY, ALPENA, MI

on the new *Bulgaria*. The article alleged that the ship came into the repair yard with many planks not fastened to every frame and that in one 40-foot-long plank, "there was only just fastening enough to keep it in place." The ship, the article asserted, would probably need an additional seven or eight tons of iron fastenings and a complete recaulking, as "in several places daylight can be seen through her ceiling and planking."[44]

In the following issue of the *Marine Record*, an enraged Davidson provided a vigorous response to correct "these outrageous and unjust statements." Following a detailed description of the ship's architecture, he closed his letter with these words:

> Building a steamer as described above, with all her improvements and her entire cost embracing a fortune, it would have been the height of folly for me to have neglected fastening a few planks by not putting in a few bolts when we are getting our iron for less than 3 cents per pound.
>
> The statements concerning the *Bulgaria* in your paper of the 19th inst. does her an injustice, and reflects upon the inspector, her owner, and her builder. I make this explanation through your columns to correct the wrong impression that your article conveys, and if you will insert this letter in your paper I will thank you very much.[45]

The source of the slanderous article is unknown—and it was slander. The editors of the *Marine Record* seemed to favor the Cleveland ship owners, and it could have been Corrigan who orchestrated the column—although plenty of competitors disliked the aggressive and unconventional Davidson. But no public feud broke out between Corrigan and Davidson over the two ships, and the *Bulgaria* and *Roumania* went on to long service lives. Never again, it appears, did Corrigan purchase a Davidson vessel. As was usual for James Corrigan, his ship and cargo were fully insured—$45,000 for the *Polynesia* and $12,000 for the lost coal.[46]

James Corrigan's life and influence on the industrializing Great Lakes is beyond the scope of this book. He became, for a time, a dominant force in industry, including serving as the president of the Lake Carriers Association—which for several years held its meetings at his Cleveland office. Corrigan, however, took financial risks and seriously overextended in building up his mining interests. In the Panic of 1893, his main company, Corrigan, Ives, & Company, went into receivership. Financially compromised but not ruined, Corrigan maintained his status as a player in the mining industry and retained his fleet (in partnership with his brother

John). In 1894, with capital from new partners, he reorganized as Corrigan, McKinney & Company and was again elected president of the Lake Carriers Association. As of 1894, his fleet included five large and relatively new wooden steamers, the *Australasia* (1884), the *Bulgaria* (1887), the *Aurora* (1887), the *Caledonia* (1888), and the *Italia* (1889); and three large, old wooden schooners: the *Tasmania* (formerly the *James Couch*) (1871), the *Northwest* (1873), and the *George W. Adams* (1874).

Survival of the Fittest

Responding to questions from a Cleveland marine reporter about the upcoming 1895 season, James Corrigan struck a Darwinian tone, predicting "the struggle of the survival of the fittest in lake navigation" between the large new steel ships carrying more than 5,000 tons and the smaller vessels carrying less than 2,000 tons.[47] It was, in reality, a struggle between those owners who had invested heavily in newer and larger steel bulk steamers and those, like him and Captain Davidson, who had not. It was not really about larger and smaller vessels but about the cost of acquiring and maintaining them. Between 1895 and 1897, James and John Corrigan ordered the construction of four steel schooner barges, the *Aurania* (1895), the *Australia* (1897), the *Amazon* (1897), and a second *Polynesia* (1897)—all of them exceeded 3,000 gross tons and were capable of carrying more than 6,000 tons of iron ore.

Had Corrigan been deliberately transitioning to a modern fleet, the argument against his character would be hard to make—it was the formative period of American industrial capitalism. However, as a former mariner and long-time ship owner, James Corrigan clearly understood the risks faced by the crews of older wooden schooners. In every instance uncovered, he seemed to have taken out the largest insurance policies that the underwriters allowed. In his version of the survival of the fittest, it mattered only that his ore was delivered, not the condition of the ship or the welfare of the mariners who delivered it.

Corrigan's Darwinian philosophy is most obvious in his acquisition of older wooden steamers and schooners for the 1899 season. He began the year with a fleet of four large wooden steamers and four steel schooner barges—all large and relatively modern vessels—and just one old vessel, the former *James Couch*, now called the *Tasmania*. During that year, he bought sixteen more wooden ships (more than any other owner on the Great Lakes that season).[48] All were older, smaller, and less expensive than the new steel vessels.

Originally named *James Couch* when built in 1871, the renamed schooner *Tasmania* became another one of James Corrigan's "death ships" when it sank with loss of eight hands in October 1905. C. PATRICK LABADIE COLLECTION / THUNDER BAY NATIONAL MARINE SANCTUARY, ALPENA, MI

During the mid-1890s, the market value of older wooden vessels had diminished significantly as the largest class of wooden and larger steel vessels proliferated and helped to push down average freights rates. In 1892, the average rate to carry iron ore from Ashland, Wisconsin, to Lake Erie was $1.20. In 1898, it was just sixty-one cents. In 1899, however, Lake Superior ore production jumped substantially, especially at the mines of the Mesabi Range, where production went up from 4.6 million tons in 1898 to 6.6 million tons. More significant for Corrigan, the price of the ore delivered to Lake Erie increased from between $2 and $3.50 per ton (depending on source and grade) to between $4 and $6.48 per ton, and pig iron, which went from $10.30 per ton in 1899 to $24.15 in 1900.[49] For Corrigan, these facts presented him with both economic necessity and opportunity. The necessity was to guarantee that he could get his valuable ore to market during a period of ship scarcity, and the opportunity was that he could make high profits while doing it. The *Marine Review* commented on the skyrocketing value of "the older class of wooden vessels" and James Corrigan's strategy of buying them up:

> It is the intention of the Review to present with the close of the year a list of transfers of lake vessels, on account of the great number of changes in ownership, especially in wooden vessels, that have taken place during the past season. The list will very probably show that James Corrigan of Cleveland (Corrigan, McKinnie & Co.) has been the principal purchaser of wooden craft. He had ore to move for which vessels of medium size were required. In the beginning it was thought that he was treading on dangerous ground in wholesale purchases of vessels of this class, but it would now seem that if he had not bought the vessels, it would have been impossible for him to move his ore. He has the ships for another year that will be practically a repetition of the one now drawing to a close, and in serving in the sale of ore as well as the carrying of it they will probably pay for themselves in the two seasons.[50]

By the end of the 1905 season, nine of the sixteen ships Corrigan purchased in 1899 had been lost in his service. One steamer, the *Minnesota*, burned after a lamp exploded in the engine room; eight vessels sank, including the schooner *Charles Foster*, which foundered with all eight hands on December 9, 1900. Eight more died on October 20, 1905, when the *Tasmania* sank, again killing the entire crew. Two of the steamers, the *Robert Wallace* and the *Iron Chief*, went down after their stern pipes (the cutlass bearing where the propeller shaft comes through the hull) failed—an obvious sign of age and poor maintenance. Four of the ships purchased in 1899 did not even make it through the two seasons mentioned in the *Marine Record*. What is especially remarkable is that James Corrigan managed in every instance to secure substantial insurance on both the hulls and cargoes, including an unusual December policy on the doomed schooner *Charles Foster*.

James Corrigan purchased one other old schooner in the fall of 1899. The 97-foot-long, 84-ton *Idler* was a famous luxury sailing yacht built in Connecticut in 1864 and a second-place finisher in the 1870 America's Cup races. The *Idler* had fallen on hard times, but over the winter, Corrigan had the schooner thoroughly rebuilt and refurbished. As captain, Corrigan engaged Charles Joseph Holmes, a mariner with a dubious record that included the loss of two men in the wrecking of the steamer *Wallula* in 1896 and a penchant for self-promotion. Holmes may have attracted Corrigan's attention through a newspaper article that appeared in June describing Holmes's intentions of sailing around the world in his 20-foot yacht. A thrill seeker and a teller of tall tales, Holmes was the wrong man for the position.

In late June, the revitalized *Idler* set sail on its first voyage. Among the passengers were James Corrigan, his wife, three daughters, and granddaughter as well as John Corrigan and his wife and daughter. The family enjoyed several days of calm weather yachting in the St. Clair Flats area before heading for Cleveland. Both of the Corrigan brothers left the yacht on the journey home, leaving their families in the care of Captain Holmes and his highly experienced crew. The morning of Saturday, July 7, found the *Idler* flying through the seas at a reported ten to twelve knots with all sails set. As the day wore on and the weather deteriorated, Captain Holmes did nothing to further prepare the schooner for heavy weather. Despite signs of a squall forming in the distance, warnings from a passing fishing tug captain, and the suggestions of the mate, Holmes refused to take in the sails until it was too late. Two squalls, one with winds estimated at sixty miles an hour, caught the sails and knocked the *Idler* over onto its side. Within minutes, water pouring in through the unsecured hatches and windows filled the vessel, sending it to the bottom with four of the women trapped inside. In the end, all of the crew members were rescued, but only one of the seven Corrigan family members—Mary, John's wife—survived the wreck.[51]

Newspapers across the country provided extensive coverage of the tragedy and its aftermath. The facts seem clear: despite his denials, Captain Holmes's reckless behavior and failure to secure hatches led directly to the sinking. Competently handled, the *Idler* could have easily weathered the blow. During the inquest, the mate testified that when asked about

Built in 1877, the Corrigan schooner *Charles Foster* foundered with eight crew in December 1900. C. PATRICK
LABADIE COLLECTION / THUNDER BAY NATIONAL MARINE SANCTUARY, ALPENA, MI

reducing sail, the captain refused and exclaimed, "Keep it on and have a little excitement."[52] After the inquest, Holmes was arraigned before a federal commissioner and bound over to a grand jury, facing a federal charge of manslaughter. Holmes was briefly jailed before posting a $1,000 bond. After months of delays and just days before his case was scheduled to go to trial, Holmes jumped bail and disappeared.[53] He was ultimately found and returned to the Cleveland jail.

While stewing in prison, Holmes leveled a series of charges against James Corrigan "concerning the management of certain vessel property." These charges were ignored by the US district attorney prosecuting the case.[54] In August 1901, Cleveland's newly elected and reform-minded Democrat mayor, Tom Johnson, and real estate dealer John Curran (strangers to Captain Holmes but not to the staunchly Republican Corrigan brothers) signed another $1,000 bond for Holmes's release. Although Captain Holmes almost certainly carried primary responsibility for the tragedy, a manslaughter charge was highly unusual in such circumstances. On February 20, 1902, Judge Francis Joseph Wing of the United States District Court in Cleveland finally dismissed the charges due to a lack of available witnesses. Holmes had paid a price for his actions, having spent several months behind bars.[55] He spent the rest of his life involved in a variety of real and invented adventures before disappearing along with 305 others on the US Navy collier *Cyclops* in April 1918.[56]

A second-place finisher in the 1870 America's Cup races, shown here, James Corrigan's sailing yacht *Idler* overturned and sank in July 1900, taking the lives of his wife, daughters, niece, and grandchild. *ILLUSTRATED LONDON NEWS,* SEPTEMBER 3, 1870

The loss of his family on the *Idler* haunted James Corrigan. While divers recovered some of the victims from the wreck, he became obsessed with recovering all the bodies—advertising extensively, offering rewards, and building a glass-bottom scow to look below the surface for human remains. One wonders what crossed Corrigan's mind during the dark days after losing most of his family. Did he ever think about the families of the men who risked their lives and those who died carrying ore on the decrepit vessels of his motley fleet? It seems not, for he continued with business as usual. Eight men died on his schooner *Charles Foster* on December 9, 1900. Five years later, eight more followed on the *Tasmania*.

The purpose of this chapter is not to vilify James Corrigan; he was a tough man in a rough age, and he paid a horrible price. Rather, the larger point is to highlight how rapid industrialization on the Great Lakes between 1880 and the mid-1900s amplified the dangers faced by mariners—especially those caught up in the iron trade under a ruthless owner. Shipwrecks were a fact of life on the Great Lakes throughout the nineteenth century, and other major ship owners during the early industrial era may have had safety records as bad as James Corrigan's. However, over the course of his career—from the first schooners purchased in the mid-1870s until his death in 1908—James Corrigan became the principal owner of approximately thirty ships, nearly two-thirds of which were lost while in his service. His was a grim record of individual vessel accidents that, when looked at as a whole, were no accident. He routinely bought old ships at the lowest possible prices and drove them hard. While the heavily insured Corrigan remained protected against maritime disaster, his men risked their lives in heavily loaded and poorly maintained craft. It was a cold and terrible approach for a man who believed in the survival of the fittest.

Chapter Ten

DAVIDSON'S GOLIATHS

Collectively, Wisconsin shipwrecks preserve the material culture and the stories of arguably the fastest period of change ever experienced by a North American maritime region, the industrialization of the Great Lakes. Wisconsin's James Davidson wrecks, however, possess special importance as a collection of work by one person, who was at the same time a master shipbuilder, an active mariner, and a successful businessman. James Davidson's three professional identities are embodied in every ship he built. Unlike most other shipbuilders of the industrial period (with the exception of his contemporary Captain Alexander McDougall, the inventor of the steel whaleback), Davidson built his major ships based on his own ideas about maritime technology and understanding of regional economics, not the whims of potential customers. In 1897, the *Detroit Free Press* commented on the captain's singular independence:

> It is his rule never to build a boat on contract. He chafes under the limitations imposed by a contract, so he builds on speculation, and the steamer, when finished, embodies his ideas of what a steamer should be.[1]

Four Davidson steamers and four schooners or schooner barges became total losses or were abandoned in Wisconsin. The locations of four large steamers, one modest sized and one enormous schooner barge, are known, and all have been studied by underwater archaeologists. While the wreck events occurred after Davidson had sold the vessels, he had owned and operated each of them for periods of time before selling them at advantageous moments. Davidson was especially successful in combining immediate and medium-term economic forecasting and pragmatic shipbuilding with a vessel management style that encouraged ship operations that were efficient as well as safe (given the standards of his day). The Wisconsin collection of Davidson vessels preserves the material remains of one man's evolving vision of the Great Lakes and nation as it industrialized during the final two decades of the nineteenth and the first decade of the twentieth centuries.

The Davidson collection is also important because each wreck has been subject to a basic level of archaeological and historical documentation, and four of the wreck sites are now listed on the National Register of Historic Places. For readers interested in the nuances of the sites and technological details, published archaeological reports are available for all of the wrecks. They provide baseline information that historians and archaeologists can draw on to explore the nuanced development of Davidson ships and their adaptations by later owners. Although they have been studied, the Davidson wrecks, as a collection, offer untapped potential for future scholars to study the final, and arguably the most creative, period of wooden shipbuilding in America as well as the complex and rapid processes that transformed the Great Lakes from a maritime region dominated by older Atlantic maritime values and practices to a regionally distinct, fully industrialized waterway. Supplemented by additional historical research, this chapter draws on the previous work by archaeologists and historians and examines the five Davidson wrecks within Wisconsin's collection of Davidson's Goliaths, a term coined by former state underwater archaeologist David J. Cooper for the vast wooden ships. The attention devoted to each ship and wreck event varies, depending on the author's level of involvement in the original studies conducted by the Wisconsin Historical Society and its many partners.

Of the eight Davidson-built vessels that were reportedly lost or abandoned in Wisconsin waters, six have been found and archaeologically documented. The documented wrecks were built between the years 1884 and 1900, the most productive years of the Davidson shipyard. The oldest vessel, the *Australasia* (built in 1884), was Davidson's fifth bulk steamer and a groundbreaker in its size, internal construction features, and towing capabilities. The newest steamer, the *Appomattox* (built in 1896), represents the climax of the Davidson bulk steamer. It remains the largest steam bulk freighter ever built from wood in the United States and possibly in the world. The final vessel is the great schooner barge *Pretoria*, the first of three monsters that became the longest schooner-rigged wooden ships in American history.

Looking at the history and the archaeological remains of Davidson's large steamers, one is struck by their durability through seasons of hard use. They were built to endure the regular stresses of the heavy machinery, cargo, and surging weight of the loaded consort vessels they towed as well as repeated instances of grounding, collision, contact with ice, and storms.

AUSTRALASIA, 1884–1896

As with most of the large Davidson vessels, the *Australasia* (1884–1896) was first designed to suit the particular specifications of Captain Davidson, an active mariner and ship owner.[2] Probably the largest wooden steamer on the Great Lakes when launched in 1884, the *Australasia* was the fourth Davidson vessel powered with a compound engine. As with all of the Davidson steamers, the *Australasia* was built to tow large vessels as well as carry cargo. The 30-inch high-pressure by 54-inch low-pressure cylinder King Iron Works engine may have been the most powerful in service on the Great Lakes in 1884.

The installation of an engine more powerful than on his previous steamers, however, suggests that Davidson envisioned an expansion of the tow barge system—or at least expected the more powerful engine to contribute to the efficiency or safety of the ship and its consorts. His focus on towing is suggested by the steam windlass and capstan installed on the *Australasia*'s stern, a Davidson first that would prove useful in towing and in dockside maneuvering of the steamer and its consorts. He also installed a three-cylinder steam deck hoister for moving hatches.

At 282 feet, the *Australasia* was Davidson's largest ship yet, some 10 feet longer and more than 200 tons larger than the *Siberia*. The breadth of 39 feet was, however, the same. The *Australasia*, at 21.2 feet, was 2 feet deeper than its predecessor. While he fine-tuned the architecture and engineering with each new vessel, the *Australasia* seems to have been something of a breakthrough. Davidson went to some lengths to describe his new transverse trussing system built of 2½-inch-diameter iron rods secured by bolts, nuts, and heavy turnbuckles. In the 1880s and 1890s, accidents such as running aground or collisions were virtually inevitable, and the *Australasia* had its share of these. The remarkable fact is that the wooden steamers proved durable and often easy to repair.

In September 1886, after operating the *Australasia* profitably for two years, James Davidson sold the steamer and its consort schooner *Polynesia* to James Corrigan for a reported $160,000. The steamer remained a key vessel in the Corrigan fleet for ten years—narrowly surviving storms such as the one that claimed its consort *Niagara* in 1887 and serious collisions in 1887, 1894, and a final one on June 22, 1896, that smashed in the bulwarks, tore away half of the pilothouse, and left a twelve-foot by four-foot hole in the ship's side.[3] Quickly repaired and returned to service, the *Australasia* soon made its final voyage.

Bound for Milwaukee with 2,200 tons of soft coal aboard, the *Australasia* cleared the Straights of Mackinaw and ran into heavy northwest winds and building seas. To avoid the weather, the captain set a course to the west where they could gain some shelter from the Door Peninsula. When nearing Baileys Harbor at about six o'clock on the evening of October 17, 1896, a fire broke out. Apparently starting underneath the boilers, the fire spread throughout the ship, reaching the fuel bunkers and cargo of coal, which ignited, causing the deck to burst into flame. Captain Robert Pringle ordered the engine to full speed and headed the steamer toward shore.

Nearing Jacksonport, the ship touched bottom, and Pringle ordered the crew to bore holes in the bottom and then to abandon ship. Captain Pringle and his sixteen-man crew made it safely ashore, where they were received by farmers, and according to a local newspaper, several of the crew managed to get "four sheets to the wind" as they watched their steamer burn to the waterline.[4]

When word reached Sturgeon Bay of the *Australasia*'s condition, Captain James Tuft mobilized the tug *John Leathem* for action. The *Door County Advocate* newspaper covered the event in detail. At about ten thirty that night, the tug found the burning ship about seven miles off shore of Jacksonport. Some of the tug's crew boarded the vessel at the stern and found, still sitting on the galley table, the *Australasia*'s dinner, which they quickly ate. Realizing the boilers of the burning ship were extremely hot, Captain Tuft moved the *John Leathem* well away from the vessel, but not before a section of the burning bulwarks collapsed onto the deck of the tug—apparently causing no damage. At about midnight, the crew managed to get the flames in the bow dampened and to secure a towline—a maneuver that they were forced to repeat seven more times when the line burned through and parted. Making matters more difficult was that the rudder was stuck in a hard-over position—making towing extremely difficult. Finally, Captain Tuft purposely sunk the *Australasia* by ramming a large hole into the side. The vessel settled in fifteen feet of water, two miles south of the north point of Whitefish Bay, about 350 yards from the beach of what is now Whitefish Dunes State Park.[5]

The fire had some suspicious elements, at least based on the newspaper accounts. The crew apparently deployed only one fire hose and not until well after the fire had been discovered. It was a drunk and unhappy *Australasia* crew that boarded the *John Leathem*, with several fights breaking out during the ride down to Sturgeon Bay culminating in a dockside scrap with "some great old slugging done for a minute or two." As with all Corrigan wrecks, the *Australasia* was insured—in this instance for at least $50,000.[6]

The owner of the *John Leathem*, the Leathem Smith Towing Company, purchased salvage rights to the *Australasia* and its cargo. Initial inspection of the wreck by divers raised hopes that the steamer could be refloated, but further damage from heavy weather ended that possibility. Despite the weather, the salvage team worked quickly and within a week removed part of the engine, anchors, cables, and several scow-loads of coal. The job of salvage proved long and dangerous, with two men seriously hurt when a block gave while they were trying to recover the boilers. The yearly ice banks built up over the wreck, ending operations for the winter and further damaging the hull and machinery. Nearly a year passed before the salvagers finally removed the boilers.[7]

The *Australasia* as an Archaeological Site

The *Australasia* is part of a collection of archaeological sites in Wisconsin that document the industrialization of wooden shipbuilding on the Great Lakes. Although seriously compromised by fire, storm, ice, and salvage, the wreck of the *Australasia* is surprisingly well preserved. The ship's substantial construction imparted great durability to a wreck that has resisted more than a century of storms, ice, and human activity. Easily accessible by boat and located in shallow water, the site of the *Australasia* has broad recreational as well as heritage value for kayaking, snorkeling, and shallow water diving.

During fieldwork conducted in 2012, archaeologists found about two-thirds of the hull buried and protected by sand. The bow section protrudes up from the bottom with broken frame sets clearly visible. The archaeologists found ample evidence of the fire, including pockets of melted brass scattered around the bow. The port side was less affected by fire but eventually broke away from the main hull and lies flat along the bottom. Much of the starboard side burned away to the waterline and remains attached in the original position.[8]

Although much encrusted by invasive mussels, exposed sections of the stern preserve evidence of the *Australasia*'s boiler room and propulsion system, including brass fragments from a shaft bearing, parts of the shaft log,

Iron cross-bracing, an essential feature of the large steam vessels built by James Davidson, is visible on the wreck of the *Australasia*. TAMARA THOMSEN, WHI IMAGE ID 140130

A diver documents the hull of the *Australasia*. TAMARA THOMSEN, WHI IMAGE ID 140129

and the main shaft bearing. Burning away layers of wood, the fire exposed many of the fastenings and iron-strapping and truss elements that gave the ship its great strength. The fire also revealed the heavy bolts driven down through the center of the planks. This edge joining was another important feature on the large wooden bulk carriers that helped to keep the sides watertight and strong. The presence of the innovative transverse trussing Davidson pioneered on the *Australasia* was not confirmed, but it probably survives as well.

Despite the sand coverage, on a clear water day the site offers an exciting and immersive look into the era of the big wooden bulk steamers. Hovering above with a kayak or snorkel, or below on scuba, one can see the large timbers that sweep out from underneath the bilge, and the shape of the burned-away bow is clearly visible sticking well up from the bottom. Further evidence of fire includes the melted blobs of brass as well as the charred planks and frames. Twisted fastenings and pieces of truss, iron plate, and engine room bulkheads testify to the ship's identity as a steamer and of the extensive salvage. A severely burned section of the port side that fell outward provides a close-up view of Davidson's iron-strapping systems and a visual sense of the massive amount of iron used to build these wooden ships. Unfortunately, sand covers the center of the bilge, likely obscuring the innovative transverse trusses in which Davidson took such pride.

CITY OF GLASGOW, 1891–1917

The *City of Glasgow* was one of four nearly identical steamers launched by James Davidson in 1891.[9] In the seven years between completing the *Australasia* and completing the *City of Glasgow*, Davidson built at least twenty-four vessels, among them several large bulk steamers. At 297 feet long by 41 feet in breadth and 20.5 feet in depth, the *Glasgow* and its sisters were not Davidson's largest steamers to date—the *George T. Hadley*, built in 1888, had a larger gross tonnage (2,073); and three vessels, the *Alex Nimick*, *Andrew McLean*, and *John Harper*, all built in 1890, were just over a foot longer. With a gross tonnage of 2,002, the *Glasgow* and its sisters, however, were a foot wider than any of the previous Davidson steamers and 2 feet wider than the *Australasia*.

During the 1880s and early 1890s, Davidson expanded his facilities significantly, and the capacity to build four major vessels simultaneously reveals the truly industrial economy of scale of wooden shipbuilding at his yard.[10] During this period, Davidson reported employing seven hundred men in the

One of the big four of 1891, *City of Glasgow* had a hull that was a foot wider than previous James Davidson steamers. C. PATRICK LABADIE COLLECTION / THUNDER BAY NATIONAL MARINE SANCTUARY, ALPENA, MI

shipyard and another three hundred as timber cutters in the large stands of oak he acquired.[11]

The economy of scale is also revealed in the reduction of variability between the succeeding generations of Davidson bulk steamers. In addition to limited changes in hull sizes, from 1888 forward, all the big Davidson steamers were driven by triple-expansion steam engines fed by twin scotch marine boilers. The *City of Glasgow* engine featured one 20-inch, one 33-inch, and one 54-inch cylinder.

Davidson immediately put the *City of Glasgow* into the ore trade, towing the new 190-foot-long Davidson-built schooners *Celtic* and *Adriatic*. On July 8, 1891, the ship suffered its first accident, a collision with a schooner at the Sault Locks. The schooner lost forward rigging, but the *Glasgow* suffered no real damage. Returning from Lake Superior two weeks later, the fully loaded *City of Glasgow* ran hard aground. Rather than calling in salvage tugs, the *Glasgow*'s captain notified James Davidson, who immediately traveled to the site to take charge. This level of personal involvement in the physical details of day-to-day operations is highly characteristic of Davidson but unusual among the other major ship owners.

Davidson built all his ships on speculation but had lined up the would-be iron king, Ferdinand Schlesinger of Milwaukee, to buy the "Big Four" ships for $150,000 each. However, when his client's fortunes dipped, Davidson kept the vessels.[12] The steamer went on to a typical career for a bulk carrier, with the occasional minor to medium accidents and a solid record of delivering cargo and towing consort vessels. In late June 1893, the "Scottish fleet," consisting of the *City of Glasgow* and the Davidson built schooner barges *Aberdeen*, *Dundee*, and *Paisley*, delivered a combined cargo of 260,000 bushels of wheat to Buffalo. This delivery was said to be the largest by one fleet in the port's history.[13] Davidson operated the *City of Glasgow* for four seasons before selling the vessel in October 1895 to the Buckeye Steamship Company for a handsome $110,000.[14]

In the late 1890s, the opening of the Poe Lock increased the minimum depth at the Sault Locks to twenty feet, allowing larger and more deeply laden ships to transit the canal. Like many owners of the largest wooden bulk steamers, the Buckeye Steamship Company raised the deck of the *City of Glasgow*, adding more than 3 feet of depth and an additional 400 gross tons—as well as substantial capacity—to the vessel. On its final trip as a steamer, the *Glasgow* delivered some 3,500 tons of soft coal into Green Bay, roughly twice the capacity of the 1874 steamer *James Davidson*.

On December 3, 1907, the *City of Glasgow* left the dock in Green Bay empty and headed for Escanaba, Michigan, to pick up a load of ore. Late in the afternoon, a fire, probably caused by burning cinders, broke out in the coal bunkers. The flames quickly consumed the upper works of the steamer, and within an hour, the crew was forced to abandon ship, leaving most of their belongings behind. The *Glasgow* eventually sank and was declared a total loss, with an insurance value of $60,000. Its position in the middle of

The *City of Glasgow* at Sturgeon Bay, Wisconsin, after its succesful salvage in 1908. C. PATRICK LABADIE COLLECTION / THUNDER BAY NATIONAL MARINE SANCTUARY, ALPENA, MI

the shipping channel into Green Bay created an immediate need for salvage, or at least wreck removal. Reminiscent of the burning of the *Australasia*, the Leathem Smith Towing and Wrecking Company purchased the salvage rights. The fire had compromised the integrity of the hull's iron truss/arch system, and it appeared that the ship might break in half because of the heavy propulsion machinery. After weeks of work hampered by bad weather, Leathem Smith managed to finally raise the wreck, on May 23, 1908.[15] Within a week, the wreckers managed to nurse the burned steamer from Green Bay into a dock at Sturgeon Bay. The local newspaper reported:

> The work of raising the Glasgow is a job that any wrecking company could well be proud of . . . the difficulty the wreckers had to contend with will be all the more appreciated when an inspection of the burned craft is made, as she was a mere shell, and the wonder is that they succeeded in keeping her together. . . . The hull of the boat is worthless and no attempt will be made to get anything out of it except the steel straps, bolts, and scrap iron.[16]

The ship retained greater value than the newspaper forecasted. The engine was almost undamaged by the fire, and a large section of the hull retained structural integrity. In 1910, Leathem Smith amputated 125 feet from the stern of the vessel forward. They converted the remaining 173 feet into a stone barge, which the company operated for about seven years. On October 6, 1917, the *City of Glasgow*, in tow along with the Davidson-built schooner *Adriatic*, broke loose from the steamer *Hansader* and ran aground in Lily Bay, north of the Sturgeon Bay ship canal. The *Adriatic* was released a week later and repaired. The *City of Glasgow*, however, blew farther into shore and defied all efforts to free it. The barge *City of Glasgow*, a large unpowered fragment of what had been one of James Davidson's Big Four steamers, was officially abandoned in 1922.

The *City of Glasgow* as an Archaeological Site

The remains of James Davidson's *City of Glasgow* rest in just six feet of water near the Lily Bay boat launch and county park. The wreck is easily accessible to kayakers and to those with mask, fins, and snorkel. As with the *Australasia* site, the *City of Glasgow* has none of the original steam machinery. However, it is the Davidson hulls that are of special interest, and more of the *City of Glasgow*'s hull's lower structure remains clear and

Remnants of the *City of Glasgow* wreck are visible under about ten feet of water in Lily Bay, about three miles north of the Sturgeon Bay Ship Canal.
TAMARA THOMSEN, WHI IMAGE ID 140409

easy to explore. The open bilge reveals all of the lower hull architecture Davidson employed to create the long steamer's hull strength. A single oak centerline keelson 24 inches wide and 18 inches deep, sandwiched between steel plates, formed the central backbone of the ship. Other builders of the large wooden steamers continued to use multiple center, rider, and sister keelsons along the centerline. But by substituting steel reinforcing plates along the keelson, Davidson achieved the desired strength as well as some improvement in net cargo capacity. Outboard of the center keelson on each side are six bilge keelsons—heavy 12-inch by 12-inch timbers that provided additional hull strength and helped to distribute the weight of the cargo along and across the bottom of the hull.

Visitors to the site should take note of the extreme regularity of the frames and keelson. By 1891, Davidson was producing wooden vessels with highly standardized materials and methods. The surviving section of the *City of Glasgow* represents the fully mature Davidson bulk steamer. Improvements came and sizes increased some, but by the time of the Big Four, all of the fundamentals of architecture and engineering were in place.

CITY OF NAPLES (RENAMED FRANK O'CONNOR), 1892–1918

James Davidson followed up the Big Four of 1891 with the Big Three of 1892.[17] The virtually identical *City of Venice*, *City of Genoa*, and *City of Naples* were the first Davidson vessels to officially exceed 300 feet. The steamers measured 301 feet in length, 42.5 in breadth, and 20.1 in depth, with tonnages varying slightly at just over 2,100 gross tons. The vessels were fast and could carry the largest cargoes of any of the wooden steamers then in service. On one of its first trips, the *City of Venice* set a new record for a wooden steamer, when drawing just 16 feet 2 inches of water, the ship delivered a load of 102,042 bushels of wheat.[18]

Shipbuilding on the Great Lakes was a boom-and-bust business. In good times, ship owners rushed to order new vessels, which inevitably led to periods of overcapacity followed by severe curtailment in the shipbuilding business. By the early 1890s, Davidson's strategy of financial independence and success in a diversified range of shipping, lumber, banking, and sugar businesses gave him near immunity to economic swings. Like John D. Rockefeller, another business college graduate, Davidson kept close track of financial details and always retained enough personal capital to protect himself as well as to take advantage of the inevitable economic downturns.

Davidson continued building even in bleak times, a practice that irritated other vessel owners. On completing the *City of Naples* in September 1892, Davidson told a *Chicago Daily Tribune* reporter that he "had been roundly abused by many vessel owners for continuing to build boats at a time when so many believed there were already too many boats afloat for business. . . . He is now building four large barges and expected to keep right along building boats, notwithstanding the protests."[19]

Davidson was a brilliant and sometimes a very lucky businessman, but where luck and ability part is difficult to determine. It is also difficult to know when the captain, who was frequently asked for his opinions by newspapers, deliberately used the public pulpit to manipulate business conditions or was simply responding to them—he did both. By February 1893, the national economy was beginning to exhibit the first signs of the Panic of 1893, an event that ushered in the worst depression in nineteenth-century American history. In an extended interview printed in the March 9, 1893, issue of the *Detroit Free Press*, the ever-optimistic Davidson predicted,

> I am now at work on six new freight vessels. Four of them will be ready
> for use by the opening of navigation. It is true that the lake tonnage is

being added to every year, but there is a demand for every vessel now afloat, and this demand is on the constant increase. This is a great and growing country. Thousands are coming to it every year, and this, combined with the natural increase in population, is developing its enormous resources at a rapid rate. . . . The coming season will, I think, be a most prosperous one. Last season returned fair profits to shippers and carriers, but this season will be ahead of it 50 percent. . . . Nobody has reason to feel blue who lives in the United States. It is the greatest country on earth, and nowhere else does a man have the same opportunities of becoming rich.

Asked about his future building wooden boats, Davidson replied: "I will continue it as long as I can obtain the necessary timber. In my opinion the wooden boat is the best and the strongest and the safest."[20]

Two days after Davidson's glowing forecast, the *Detroit Free Press* announced "the largest transfer of lake vessel property in recent years," his sale of the *City of Naples*, *City of Genoa*, and *City of Berlin* to the Gilchrist Transportation Company for a total price of $405,000. In fact, Davidson had been shedding vessels since the beginning of the year, selling two large schooners for a combined $75,000 and the *City of Paris* for an estimated $140,000. By the summer of 1893, the stock market was in freefall, with banks and businesses beginning to collapse around the country. Flush with capital, James Davidson just kept on building ships, now taking advantage of the depression-reduced price of materials and wages.[21]

The Gilchrist family left detailed records of their shipping enterprises, including the acquisition and ownership of the Davidson vessels. Davidson not only built the ships but also provided long-term financing for them at a steep 6 percent annual interest to buyers, including J. C. Gilchrist. Gilchrist did well in the 1890s. For example, in 1896, the Gilchrist fleet of twenty-three vessels earned $281,937.84 in freight and towing revenue, with the company paying out $66,920 in dividends to stockholders. At the end of the 1898 season, Gilchrist purchased four older wooden bulk steamers for a mere $70,000, a bargain one observer suggested finally made up for the "biggest mistake Gilchrist ever made," his high-priced, 6 percent interest purchase of the Davidson steamers.[22]

J. C. Gilchrist did well with the new steamers, and with the new century, he emerged as one of the largest independent vessel owners on the Great Lakes. Regardless of their original price and terms, the Davidson ships remained a good investment for Gilchrist and he continued to invest substantial capital in them including raising the deck and sides of the *City of Naples* 18 inches, increasing the ship's capacity by about one-third. During the boom year of 1900, the *City of Naples* returned a $38,000 dividend on total freight earnings of $75,895.15. This was no fluke, as the *City of Genoa* returned $28,000, and Gilchrist's recently acquired Davidson steamer *John Harper* (built in 1890) returned a whopping $47,000 dividend.[23]

The big years emboldened Gilchrist to borrow heavily and plough money into becoming one of the largest ship operators on the Great Lakes. Debts slowly mounted. In 1902, the Gilchrist Transportation Company reported more than $1 million in receipts for freight and towing but distributed just $50,000 in company dividends. A sister company controlled by J. C. Gilchrist, the Steel Steamship Company, reported more than $440,000 in receipts, but after paying $155,000 in debt service, including $140,000 to the American Shipbuilding Company, it returned no dividends. Granted, it was not a good year—even so, the *City of Genoa* produced more than $10,000 in profits that season, while the *City of Naples* suffered significant damages and managed to break even. The following year, the two steamers returned a combined profit (after all expenses and repairs) of more than $25,000.[24]

In 1903, the Gilchrist fleet lost six wooden steamers and one schooner through a variety of mishaps. These were perhaps the first in a line of business and personal disasters that brought about the company's downfall. In 1906, the beginning of another national depression hit the company hard, and in 1907, Gilchrist suffered a disabling stroke that accelerated the company's decline into receivership, which occurred in 1910. The remaining Gilchrist fleet was sold off in 1913, with the *City of Naples* going to Norris & Company of Chicago. In 1914, the steamer was sold to the Tonawanda Iron and Steel Company. In 1915, with the economy stagnant, the *City of Naples*—an old steamer of minimal value—remained tied to the dock.

In 1916, Tonawanda ship chandler James O'Connor bought the *City of Naples* and renamed it the *Frank O'Connor* in honor of his son. America's entry into World War I made the purchase a sound investment. With coal rates at one dollar or more per ton, lumber freights up to four dollars per thousand board feet, and an insatiable demand for Lake Superior wheat, old wooden steamers such as his could generate a lot of money in a short time. O'Connor got almost three seasons out of the ship before disaster struck.

On October 2, 1919, the *Frank O'Connor* entered Lake Michigan, bound for Milwaukee with 3,000 tons of coal. With the weather good, the captain set a course down the west side of Lake Michigan. At about four o'clock in the afternoon, the crew discovered a serious fire, and Captain William Hayes, the owner's son-in-law, turned the steamer toward shore. About two miles from Cana Island, the steering apparatus burned away. With the fire roaring, the steamer doomed, and land close by, the crew abandoned ship. The Cana Island lighthouse keeper and his assistant, who observed the billowing smoke, came out with a powerboat and took the retreating crew in tow. The old ship, one of James Davidson's Big Three of 1892, burned well into the night, until finally sinking in sixty to seventy feet of water.

The cause of the fire was never determined, but the fire was believed to have started in the bow. Once the blaze ignited, the caked-on residue from cargoes of grain and coal aided by the decades of paint covering the well-seasoned wood of the cabin and bulkheads provided ample fuel. The US steamboat inspectors looked into the loss of the *Frank O'Conner* and concluded that the captain was blameless.

The ship was old, but the anthracite coal was valuable and prompted unsuccessful efforts to locate the wreck. It was not until 1923 that two local men, using a thousand-foot-long rope suspended between two powerboats, found the steamer. Subsequent salvage efforts recovered less than a quarter of the original 3,000-ton cargo. After a final unsuccessful salvage attempt in the 1930s, the location of the *Frank O'Connor* was forgotten.

The *City of Naples/Frank O'Connor* as an Archaeological Site

In the late 1980s, Chicago wreck hunters relocated the site of the *Frank O'Connor*.[25] Concern for the site and its artifacts led other conscientious sport divers to report the discovery to the Wisconsin Historical Society in the fall of 1990. A non-disturbance underwater archaeological survey followed in the summer of 1991.

Today, because of the effects of invasive mussels, divers on the *Frank O'Connor* can experience one hundred feet or more of underwater visibility. When the site was originally surveyed by four archaeologists during a seven-day project, the typical visibility was between ten and twenty feet, making the mapping of the 300-foot burned steamer an extraordinary challenge.

The site is a genuine underwater museum where divers can access the massive wooden structures and machinery that characterized a fully developed Davidson wood bulk steamer. The burning away of the vessel's sides exposed much of the hull architecture and engineering but did little damage to the propulsion systems. Major hull features, such as the steel-plate-sandwiched 20-inch by 20-inch centerline keelson, 12½-inch-thick by 12-inch-wide floor keelsons, and massive triple-framed floor timber, give a clear sense of the architecture that gave these ships such durability. Although the fire probably started in the bow and that section is heavily burned, the area preserves many in situ artifacts, including a Trotman bower anchor, a mushroom anchor, the steam windlass, and a steam radiator.

The intact propulsion machinery is what makes the *Frank O'Connor* truly distinctive among Wisconsin's Davidson shipwrecks. The 20-foot tall, three-cylinder steam engine and massive twin scotch marine boilers dominate the site visually. The wooden deck below the boilers collapsed during the fire, dropping them roughly ten feet onto the bilge but causing little damage other than to move them a few feet off of the vessel's center. The *City of Naples* and its two sisters, the *City of Venice* and the *City of Genoa*, were Davidson's Big Three of 1892 and were known for their speed. Although contemporary sources report the *Naples*'s triple-expansion engine having 20-inch, 33-inch, and 54-inch cylinders (the same size as the *City of Glasgow*'s cylinders)—archaeological documentation of the engine in 1991 by the Wisconsin Historical Society found 20-inch, 41-inch, and 57-inch cylinders, evidence of a substantially more powerful engine than the historical record suggests. A 10-inch diameter shaft connected the engine to the four-bladed propeller, which was 12 feet 1 inch in diameter.

Lying flat on the bottom adjacent to the propeller is the 27-foot 2-inch rudder, complete with the steering quadrant and steering chain, consisting of 4-inch-long links. Steering systems were a persistent vulnerable point in the giant wooden steamers and schooners, and the *Frank O'Connor* proved no exception. Hovering above the rudder and near the propeller provides the diving visitor with a visceral sense of the enormous size of Davidson's wooden bulk carriers and of the heavy stresses they endured while carrying industrial cargo on the Great Lakes.

THE STEAMER *APPOMATTOX*, 1896–1905

The American economy tanked in 1893, but everyone knew it would come roaring back. The question was, how long would it take? After the Panic of 1873, it took more than six years to see a full revival. In a time of wildcat investing in mines, mills, and ships, few people among the Great Lakes shipping community had the resources necessary to play a waiting game. For shipbuilders, the situation was dire—unless your name was James Davidson. When his West Bay City rival Frank Wheeler was cutting wages and laying off men at his yard, Davidson was putting the finishing touches on another bulk steamer called the *Thomas Cranage* that he launched in July 1893.

Archaeological site plan of the *Frank O'Connor*

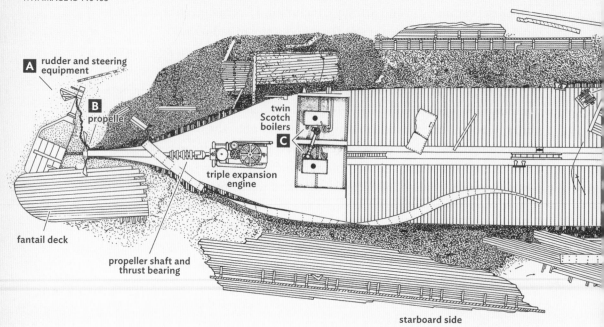

A rudder and steering equipment

B propeller

twin Scotch boilers

C

triple expansion engine

fantail deck

propeller shaft and thrust bearing

starboard side

A. The rudder and steering quadrant of the Frank O'Connor.
TAMARA THOMSEN, WHI IMAGE ID 140406

B. The propeller that drove the *Frank O'Connor* remains connected to the shaft and the triple expansion steam engine. TAMARA THOMSEN, WHI IMAGE ID 140407

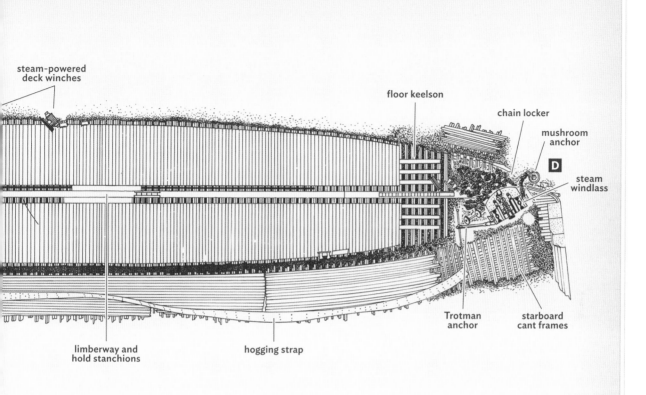

steam-powered deck winches

floor keelson

chain locker

mushroom anchor

D

steam windlass

Trotman anchor

starboard cant frames

limberway and hold stanchions

hogging strap

C. Twin Scotch marine boilers powered the *Frank O'Connor*'s triple expansion steam engine. TAMARA THOMSEN, WHI IMAGE ID 140404

D. The anchor and windlass survived the fire that started in the bow of the *Frank O'Connor*. TAMARA THOMSEN, WHI IMAGE ID 140403

Davidson's *Appomattox* was the largest wooden bulk freight steamer ever built in the United States. RICHARD J. WRIGHT, BOWLING GREEN STATE UNIVERSITY

At 305 feet in length, 43 feet in breadth, 20.7 in depth, and 2,219 gross tons, the *Thomas Cranage* was only slightly larger in dimensions than the Big Three vessels and had the same sized triple-expansion engine, but it was touted by newspapers across the Midwest as the largest wooden boat in the world.

Two weeks later, after completing the *Thomas Cranage*, Davidson tried to quash rumors that he had new ships planned and indicated that there would be no more until the economy improved. On November 20, 1893, Davidson provided a very different story to the *Detroit Free Press*.

> We're going right ahead building boats just the same as if the times were good. I am not at all scared over bad business for one season. I have kept pegging right along on that line, good times and bad, and I find that vessels average up as big profits as any other line of investment a man can find.[26]

Davidson continued to build ships throughout the darkest days of the depression. As steamers continued to get incrementally larger and to tow larger consorts, Davidson began producing fully rigged wooden schooner barges of unprecedented sizes. He also experimented with new types of vessels.

In 1895, Davidson built two huge wooden barges for the Lake Michigan Car Ferry Transportation Company. Car Ferries No. 1 and No. 2 were 317 feet in length and had a capacity for twenty-eight freight cars on their open decks. Costing about $48,000 each and equipped with steam steering and towing gear, the barges performed very well. The overall ferry service, however, was plagued by disasters, and no more barges were ordered.[27] Davidson also built two smaller barges in 1895, the 147-foot-long *Tycoon* and *Mikado* that were sold in 1897 to the Michigan and Ohio Car Ferry Company, for use as car ferries between Toledo and Detroit.[28]

By the mid-1890s, the large steel ship had supplanted the wooden ship in the public's imagination and in the economic calculus of many of the principal ship owners on the Great Lakes. Maritime pundits predicted that the coming of the anticipated twenty-foot navigation channel would mark the demise of the modest-sized wooden bulk carrier on the Great Lakes. Although fully aware of the long-term outlook in shipping, Davidson understood—correctly—that the time of the large wooden ship had not yet passed. A large, thoroughly modern steamer paired with a massive, unpowered wooden schooner barge, when well-managed, could compete with—and sometimes out-compete—the larger and more expensive steel ships coming out of the shipyards. It was on this niche that he focused his most considerable energies as builder and fleet operator. Between 1894 and

1903, Davidson built ten steam bulk carriers and 14 schooner barges, with all but one joining the Davidson Steamship Company fleet. During this period, when the demand for wooden bulk carriers was weak, Davidson became the operator of the largest fleet of wooden ships on the Great Lakes, and for a time, regional newspapers identified him as the largest independent ship owner in the United States.

By the 1890s, James Davidson was self-insured; he owned his own tugs, a well-equipped wooden shipyard, and large stands of uncut ship timber. Such assets, coupled with his intensive approach to management, allowed the Davidson Steamship Company to eke out profits even when the freight rates were low and to provide fabulous returns during flush times.

An article in *Cassier's Magazine*, written by Joseph R. Oldham, a senior naval architect from Great Britain then living in Cleveland, captured the state of Great Lakes shipping and shipbuilding during the year of the *Appomattox*'s launch. Oldham, a thirty-year veteran in the repair of iron and steel ships, was chiefly interested in the remarkable expansion in the size and efficiency of Great Lakes ships, but he made important observations germane to the Davidson fleet. The smaller vessels, Oldham explained, could load cargo efficiently at the many older and smaller docks. More important in terms of the bottom line, Oldham reported that a steamer and barge combination (such as the *Appomattox* and its various consorts) required just two-thirds of the effective horsepower needed by a larger single vessel to carry the same volume of cargo.[29] During periods of fierce competition, when the smallest operational expense margins could make the difference between large profits and major losses, Davidson vessels were less expensive than the new steel steamers to build, maintain, and operate, at least when under his management.

When construction began on the *Appomattox* in 1895, economic conditions on the Great Lakes had improved with iron ore shipping rates from Ashland, Wisconsin, hitting $1.70 per ton and Duluth wheat approaching five cents per bushel. These rates prompted the *Detroit Free Press* to forecast a long and dangerous conclusion to the season, with owners keeping ships in operation "just as long as the elements, principally ice, will let her."[30]

Taking advantage of the moment, but looking ahead to the 1896 season, the canny Davidson contracted to carry 100,000 tons of ore from Duluth to Lake Erie ports at one dollar per ton. On September 21, 1895, the *Chicago Inter Ocean* reported that James Davidson would have six large new wooden vessels completed for the 1896 season. Holding up his success as "another evidence of American life," the paper provided a brief but telling overview of Davidson's public persona:

Captain Davidson is one of the most peculiar figures among lake vessel-men . . . he has taken all the chances, and won, now being reputed to be worth between $2,000,000 and $3,000,000. The panic of 1893 did not change his plans at all. He kept on building boats, taking advantages of the low price of material and still lower wages. Many vesselmen thought that Captain Davidson had lost his head, and in some respects he was the laughing stock among the more conservative vessel owners. . . . Now that the boom in lake traffic, which he so confidently predicted a year ago, has come, Captain Davidson will see out his fleet at a profit of not less than $300,000, if not a full $400,000.[31]

At the close of the season, the *Chicago Inter Ocean* announced that a "season which began in gloom and despondency ended with high spirits and bright hopes for the future." By October, modern boats were generating seasonal investment profits of 45 percent, sparking "the greatest boom shipbuilding has ever had."[32]

Given these conditions, Davidson chose to keep his fleet and went into the 1896 season with twenty-one ships in service.[33] During the spring and early summer, he added three identical new 2,040-gross-ton Davidson-built schooner barges: the *Abyssinia*, *Algeria*, and *Armenia*.

On July 25, 1896, Davidson launched the *Appomattox*. At 320 feet in length, 42 feet in breadth, a critical 23 feet 4 inches in depth (more than 2 feet deeper than any of his previous steamers), and 2,643 gross tons, the *Appomattox* was appreciably larger than the previous generation of Davidson steamers—the *Rappahannock*, *Sacramento*, and *Shenandoah*, built in 1894 and 1895—308-foot-long vessels that measured just under 2,400 gross tons.

For the largest wooden steam freighter ever built, the event received little attention outside of the Saginaw Valley. Whereas just a year or two before, newspapers across the Great Lakes often devoted several column inches to describing a new Davidson ship, the only widely circulated notice of the new *Appomattox*'s launching was about the violent shifting and collapse of the steamer's tall smokestack.[34]

When the *Appomattox* went into service, the wild enthusiasm of the previous winter had degenerated to despair with freight rates below the grim opening months of 1895.[35] The falling rates prompted Davidson to slow the pace but not to stop work at his shipyard. In October, he announced the building of a new series of vessels and purchased 2.2 million board feet of Norway pine—enough for the decks and cabins of several large ships—but he would never again build a steamer the size of the *Appomattox*.[36]

The Great Lakes shipping community entered 1897 prepared for another disastrous season. At the annual meeting of the Lake Carriers Association, reporters observed that "every move in the proceedings was toward a reduction of expenses, directly or indirectly, in order to prepare for the survival of the fittest." With the twenty-foot navigation channel poised to open, the owners of smaller ships were bracing for economic extinction, believing that they could not compete with the new 6,000-ton steel ships under construction.[37]

Davidson began the 1897 season with eighteen large and modern wooden ships. During the year, he added the giant schooner barge *Crete* and the 263-foot *Venezuela*, a small bulk steamer that he intended to use to move into the Lake Ontario trade. Davidson's *Appomattox* and his other large steamers and mammoth schooner barges not only survived the 1897 season but thrived. At one point in August, the Davidson Steamship Company had two independent tows, consisting of a large steamer and multiple schooner barges crossing the lakes, carrying an aggregate of nearly 1 million bushels (nearly 28,000 tons) of grain.[38] He worked the fleet hard and, as was often the case, his ships made some of the last trips of the Great Lakes season, a benefit of Davidson's policy of self-insurance. Most other carriers either could not get insurance after December 1 or had to pay very substantial premiums in order to sail.[39]

The Davidson Steamship Company operated the *Appomattox* throughout the 1898 and most of the 1899 seasons. The average large steel vessels in 1898, according to the *Detroit Free Press*, were returning dividends of between 5 and 12.5 percent on investment, while large wooden vessels were paying between 4 and 6 percent. Davidson did not report his profits, but they were certainly at least on the high end of the estimates for wooden vessels and probably more in line with or surpassing those of the steel ships. During the 1898 season the Gilchrist Transportation Company paid out a 10 percent dividend on the *City of Naples* and 12 percent on the *City of Genoa*, and Davidson was a better fleet manager. Davidson's luck in avoiding serious disasters was a source of comment among the shipping circles. The *Detroit Free Press* noted that most other owners who had gone uninsured, even large operators such as Cleveland's M. A. Hanna, opted for insurance after losing a ship. The article closed with this comment: "Will Capt. Davidson follow the rule? He seldom does."[40]

As the stories of James Corrigan and J. C. Gilchrist revealed, 1899 was a banner year for the owners of wooden ships. The spiking demand for iron and other commodities outstripped the collective capacities of the fleet owners, and Davidson began to get offers for his vessels that he could not refuse. On February 4, 1899, he sold the large schooner barge *Chickamauga* for $70,000 cash.[41] At the beginning of April, the identical schooner *Chattanooga* went to the Cleveland Cliffs Iron Company for $70,000.[42] In August, Davidson sold the *Appomattox* along with the 324-foot-long Davidson-built schooner barge *Santiago* to the Boston Coal Dock and Wharf Company, a New Jersey–based firm that operated out of Duluth, for $230,000.[43] Owned by Boston Coal and Wharf and managed by Pickands, Mather, and Company of Cleveland, the *Appomattox* and *Santiago* continued to carry coal, ore, and the occasional cargoes of grain for the next six years.

Even with large ships of the most modern construction, the consort system remained vulnerable to accident. In early August of 1900, the *Appomattox–Santiago* team was involved in a serious incident. Traveling in light fog on the St. Clair River, the *Santiago* veered off course and smashed into and sank the schooner barge *Fontana*, taking the life of one of its crew.[44]

On November 2, 1905, poor visibility contributed to another accident involving the pair. Proceeding southward along the west shore of Lake Michigan with loads of coal, the *Appomattox* and *Santiago* encountered a dark bank of dense industrial smoke. This combination of fog and industrial smoke obscured the range lights on the north end of Milwaukee. The vessels' nighttime course brought them in too close to shore, and they, along with a third vessel, the bulk carrier *Iowa*, went aground on a rocky bottom near North Point. Immediate assistance by tugs, a Revenue Service cutter, and crews from the US Lifesaving Service Station led to the quick release of the *Santiago* and *Iowa*. The *Appomattox*, however, had gone the hardest aground and suffered considerable bottom damage. During the morning, the weather worsened substantially and the sea state—according to the life-saving station log—went from moderate to high. This weather compounded the vessel's injuries. For the next thirteen days, the wrecking crews, assisted by the US Lifesaving Service, fought to salvage the vessel. The damage, how-ever, proved too severe. The *Appomattox*'s bottom had broken in several places, and despite several pumps, the vessel would not float. The wreckers abandoned it on November 15, 1905.[45]

The *Appomattox* as an Archaeological Site

Today the broken remains of the largest wooden cargo steamer built in the United States rests off of Atwater Beach in Shorewood, Wisconsin. Located in fifteen to twenty feet of water, the wreck site has been flattened by more than a century of ice and waves as well as the salvaging of its machinery by the famous Reid Wrecking Company in 1907. The hull of the *Appomattox* is more intact than the *City of Glasgow*, but it lacks the intact engineering features of the *Frank O'Connor*.

As Davidson's largest bulk steamer, the surviving hull architecture of the *Appomattox* is of central interest. As with all of the large Davidson steam-ers, the bottom of the ship along most of its length consists of heavy triple frames, in this case 16 inches deep (molded) by about 13.5 to 16 wide (sided). Atop the frames is a 17-inch-wide by 18-inch-deep oak keelson sandwiched by two 1-inch-thick by 17-inch-deep steel plates. That the two earlier David-son steamers had somewhat larger keelsons suggests that Davidson increas-ingly relied on the lighter but stronger steel to provide longitudinal support.

Although the site is sometimes exposed to rough weather, a factor that complicated salvage efforts, divers can safely access the wreck through a sea-sonal Wisconsin Historical Society mooring buoy. On many days, the wreck site is clearly visible from the surface, making it an excellent destination for kayak enthusiasts and the more adventurous snorkeler. The detailed site plan created by Wisconsin Historical Society underwater archaeologists during a 2003 survey provides an excellent guide to about 250 feet of the bottom of the *Appomattox* as well as about 260 feet of the vessel's port side. During projects undertaken in 2011 and 2014, archaeologists located and documented the remaining areas of the wreck.

An archaeologist investigates heavy floor keelsons near the bow of the *Appomattox*. Running the length of the ship, the multiple keelsons helped to strengthen the long hull. TAMARA THOMSEN, WHI IMAGE ID 140115

A diver investigates a propeller of the *Appomattox*. WHI IMAGE ID 122855

A critical innovation of shipbuilder James Davidson was his buttressing the central keelson with iron keelson plates, as seen here in an underwater view of the *Appomattox*. Lighter and stronger than wood, the iron keelson plates reduced overall ship weight and increased hull strength and cargo capacity. TAMARA THOMSEN, WHI IMAGE ID 140117

Massive wooden knees connected the *Appomattox* hull frames to heavy deck timbers. The deck timbers tore away as the wreck came apart but exposed some of the fastenings. TAMARA THOMSEN, WHI IMAGE ID 140402

The *Pretoria* and the Great Schooner Barges, 1900–1905

James Davidson's final major shipbuilding achievement was to build the longest schooner-rigged wooden freight carriers in the world. While he seems to have taken great pride in building the large "marine monster" wooden bulk steamers during the 1880s, by the time he began building his largest steamers and great schooner barges in the mid-1890s, the focus was more directly about business. Davidson understood as well as anybody that the depletion of forests would bring the wooden age on the Great Lakes to a close, and he designed his last series of vessels to take final advantage of his remaining timber resources, labor pool, and shipyard facilities.[46]

James Davidson launched the schooner barge *Pretoria* on July 26, 1900. About one thousand people watched as the largest wooden ship yet built in America splashed into the Saginaw River; it was a respectable crowd, but not the four or five thousand who celebrated the important Davidson launchings a generation before. The season had gotten off to a slow start with coal freights soft and Great Lakes lumber producers banding together to force lower shipping rates on timber. Tugboat men at Duluth were giving ship owners grief, demanding the charter of two tugs to move the large Davidson

The launch of the *Pretoria* on July 26, 1900, at the Davidson shipyard in West Bay City, Michigan. C. PATRICK LABADIE COLLECTION / THUNDER BAY NATIONAL MARINE SANCTUARY, ALPENA, MI

vessels.[47] The tugboat men, the *Detroit Free Press* observed, made a terrible blunder: "Had they struck an owner without the Davidson backbone they might have succeeded in establishing a precedent." Between Davidson's temperament and his stock in the powerful "tug trust"—the Great Lakes Towing Company—the Licensed Tugmen's Association of the Great Lakes backed down fast.[48] At the Davidson shipyard, workers were completing the expansion of his dry dock to 440 by 60 feet, large enough for his biggest ships but too small for any thought of Davidson moving into steel ship construction—a persistent rumor throughout the late 1890s.

There was money to be made. During the 1900 season, some of J. C. Gilchrist's Davidson-built steamers returned 40 percent of their original purchase price in profits. Other wooden ship owners, such as James Corrigan, were doing equally well or better. The smoothly operating Davidson Steamship Company was probably doing equally well. The high value of wooden ships continued to hold with one of Davidson's old Big Three steamers from 1891 fetching more than $100,000, and this prompted him to begin selling off parts of his fleet. Shortly after completing the *Pretoria* in July 1900, Davidson sold the 2,000-gross-ton schooner barge *Algeria* (built in 1896) for $75,000.[49]

Despite the flush times, Davidson seemed uncharacteristically discouraged in the fall of 1900, telling a reporter that Great Lakes tonnage was now one-third too large for the present commerce.[50] Furthermore, it had been a tough year on the lakes for Davidson, with the schooner barge *Paisley* nearly going down after being caught in two November storms while in the tow of the Davidson steamer *Orinoco*. The *Orinoco* had just returned to service after suffering extensive damage in a fire at Toledo, Ohio.[51] Although it was owned by Minch Transit Company, the loss with all hands of the Davidson-built schooner *Dundee* on September 12, 1900, may have affected him as well. The same storm, one of the worst to hit Lake Erie in years, also claimed the Gilchrist steamer *John B. Lyon* and its crew of fourteen.[52]

THE WRECKING OF THE *PRETORIA*

The *Pretoria* provided solid service to the Davidson Steamship Company for more than five years. Usually towed by the Davidson steamer *Venezuela* or the *Rappahannock*, the *Pretoria* moved coal, iron ore, and occasional cargoes of grain to and from southern Lake Superior. It experienced the usual minor accidents, the slight collisions or stranding of either the *Pretoria* or its consort. At the beginning of the 1902 season, the *Pretoria* remained stranded in the Ashtabula harbor for eleven days, and ran aground at least once during the following seasons.[53] Now more than sixty years old, Captain Davidson continued to take personal charge over every serious salvage operation (a practice he continued to beyond the age of seventy). At the end of July 1905, the *Pretoria* again ran aground near Marine City, Michigan, carrying a cargo of coal, but it was released undamaged.

On September 1, 1905, the *Pretoria* and *Venezuela* loaded iron ore at the Allouez docks at Superior, Wisconsin. Among other vessels taking on cargo at the Allouez that day was the *Sevona*, a recently enlarged steel bulk carrier built in West Bay City by Frank Wheeler in 1890. The weather in Duluth had been miserable for early September, with heavy rain and the barometer falling. The US Weather Bureau morning map showed a tight band of low pressure passing over Des Moines, Iowa, and predicted fresh winds shifting to the Northeast, followed by clearing weather. Not perfect, but not really threatening either. The *Pretoria* and *Venezuela* headed for south Chicago, followed a few hours later by the *Sevona*. On Friday evening, however, the weather had deteriorated, and by the early hours of the following morning, every man and woman on Lake Superior was fighting for their lives.

At 2:00 a.m., the 370-foot-long *Sevona* was taking heavy seas over the bow and decks, about ten miles or so northeast of Sand Island in the Apostle Islands. With weather worsening, Captain D. S. MacDonald turned his vessel about and headed for shelter inside the Apostles. Moving at half speed through the darkness, fog, and rain, the *Sevona* ran hard aground on Sand Island Reef at 5:45 a.m. with the force of the collision tossing passengers and crew onto the deck. The collision also broke the ship in two, trapping the captain, mates, wheelmen, and watchmen in the bow section. They had no means of escape. The forward lifeboat had been removed during the recent rebuild and had not been replaced.

The *Sevona* was carrying twenty-four people, including crew and guests. A total of seventeen survived after enduring a horrifying fifteen-mile-long lifeboat ride into the beach at Little Sand Bay. Their survival was something of a miracle, as all of the steamer's professional sailors were trapped in the bow and the lifeboats were in the charge of engineers, oilers, coal passers, and galley staff. Passengers credited Second Engineer Adam Fiden for their survival; he had taken charge and had proved an expert at sculling. Back on the ship, with the forward section rapidly disintegrating, the trapped men put together a makeshift raft but drowned trying to make it ashore. By the time the storm abated, everything forward of the hull break on the *Sevona* had been consumed by the lake.[54]

The Frank Wheeler–built steel bulk freight steamer *Sevona* wrecked in the Apostle Islands during the same September 1905 storm that claimed the *Pretoria*. COURTESY OF THE WHS MARITIME PRESERVATION AND ARCHAEOLOGY PROGRAM

The steel frames and hull plating of the *Sevona* are well-preserved in the waters of the Apostle Islands.
TAMARA THOMSEN/WHI IMAGE ID 140416

For a time, the *Venezuela* and *Pretoria* fared better than the *Sevona* and were riding out the storm about thirty miles northeast of Outer Island. At about 7:30 a.m., the *Pretoria*'s "improved hydraulic steerer," the most modern available in 1900, failed. Alerted of the situation, the captain of the *Venezuela* attempted the difficult maneuver of turning his storm-tossed ship around with the disabled consort in tow. The intention, as with the *Sevona*, was to seek shelter in the Apostles. The surging weight of nearly 8,000 tons of ore plus that of the two big ships proved too much, and the towline parted at both ends. The *Pretoria* quickly drifted out of sight into the snow. The *Venezuela* searched unsuccessfully for the *Pretoria* for some time before seeking safety in the harbor of Ashland, Wisconsin.

With thirty miles of sea room, the crew of the *Venezuela* had little concern for the safety of the *Pretoria*. They had every reason to believe that the *Pretoria*—practically new, very strongly built, and immense—would weather the storm. Such things had happened before, as in the near wrecking of the smaller Davidson schooner barge *Paisley*, and had turned out fine. Furthermore, in his forty-year career as a ship owner, James Davidson had yet to lose to a storm a ship that he had both built and managed.

Although designed for independent operation in emergencies, the *Pretoria*'s disabled steering rendered it virtually powerless against the mounting

Pieces of the hull and machinery from the *Sevona* are scattered across the rocky bottom where the ship struck.
TAMARA THOMSEN/WHI IMAGE ID 140417

storm. A single sail, raised on the foremast in hopes of pulling the ship into better position to ride the waves, shredded almost instantly, which left the *Pretoria* wallowing in the trough of the heavy seas. Similar to the storm that claimed the *Lucerne* in 1886, the September storm of 1905 was a killer, one far beyond normal in its power.

The fierce northeast winds pushed the giant schooner sideways across Lake Superior at an astonishing three to four miles per hour. No large ship, not even a steel-reinforced Davidson goliath, is engineered to absorb those types of stresses. Waves crashed hard, tearing at bulwarks and bulkheads and attacking the integrity of the decks and hatches. The ship began taking on water, but the crew kept it at bay with steam pumps. After the pumps gave out, probably through drowning of the donkey boiler by a heavy sea, the schooner began to slowly lose ground. Reports from Captain Charles Smart suggest that the *Pretoria* was slowly ripped apart. Although in deep water, Captain Smart had ordered the large anchor dropped. Finally, at around half past three in the afternoon, about a mile and half northeast of Outer Island, the anchor grabbed the bottom to allow the ship to swing about with its bow into the weather.

At anchor, the disintegration continued unabated. Cargo hatches came off, and the iron-packed hold slowly filled with water. After more than an hour of punishment, the twisting stresses caused by the ship's heavy rolling in combination with the waves caused the covering board, the long cap of wood located at the junction of the deck and the side of the hull, to begin giving way, and sections of the deck started to come apart. At that point, Captain Smart and his nine-man crew launched their lifeboat and began a desperate attempt to make the beach at Outer Island. They nearly made it, but nearing the shore, the lifeboat either broached or capsized, tossing the men into churning surf. Five men drowned, but five others, including Captain Smart, were saved when sixty-year-old Outer Island lighthouse keeper John Irvine braved the seas in what the *Duluth News Tribune* called "an almost superhuman effort" and pulled them ashore.[55]

The storm was the worst in many years and claimed thirty-nine lives. Other total losses included the steamer *Iosco* and its schooner consort *Olive Jennett*, both wooden vessels built at the Frank Wheeler yard. They went down with all twenty-four of the crew.[56]

Further storms followed. A storm that struck on October 18 and 19 claimed at least sixteen ships and more than twenty lives (including the Corrigan schooner *Tasmania* with all hands). The famous Mataafa storm that occurred November 28–30 (named for the highly visible wreck of the steamer *Mataafa* at Duluth) wrecked or seriously damaged approximately forty more vessels across the Great Lakes, killing at least thirty-six mariners, including the entire nineteen-person crew of the steel ore carrier *Ira Owen*, which disappeared somewhere in the vicinity of Outer Island. The storm was particularly noteworthy as the first to take a heavy toll on the newer steel ships, which accounted for more than $3 million in losses.[57] If the Mataafa storm was a maritime disaster worthy of an industrial age (US Steel Corporation lost nineteen steel ships), then the September storm that claimed the *Pretoria* and *Sevona* was its precursor. Both events brutally demonstrated that no ship, whether built of wood or steel, towed, sailed, or self-propelled, could be counted on to face the full fury of a Great Lakes storm.

As Davidson told a reporter in Chicago, "I have owned boats for forty-five years and the *Pretoria* is the first I have lost through storms."[58] It was an extraordinary record, and the sinking of the *Pretoria* came as a shock to him. As Davidson predicted at the time, the *Pretoria* proved a total loss. The hull's structural integrity had been seriously compromised during the wreck. The washing away of the large sections of deck and deck beams placed great stresses on the sides of the heavily loaded vessel. At some point, perhaps during attempts to salvage the ore, or from the ice Davidson expected to grind up his ship, the hull broke apart, with the starboard side collapsing down onto the bilge and about half of the port side falling outboard, exposing interior architecture.

The *Pretoria* as an Archaeological Site

No site better conveys the magnitude of James Davidson's achievements as the builder of the largest wooden ships on the Great Lakes. Despite the vessel's great size and good loran coordinates, finding the wreck of the great ship proved challenging for the underwater archaeologists. Although little more than a mile from Outer Island, the location of the *Pretoria* can be a lonely and frightening place. The tip of Outer Island extends more than twenty miles out into Lake Superior, about thirty miles from the narrow entrance to Chequamegon Bay. The endless horizon and shape of the swells even on a relatively calm day hint at a smoldering power not to be trifled with.

The specifics of the keelsons, the heavy triple-framed floor timbers, and sweep belts of steel and cross-bracing do little to convey the visual power of the wrecked *Pretoria*. The bottom of the *Pretoria* is about as long as a football field, nearly all lined with almost perfectly milled 20-foot-long by 14-inch-wide ceiling planking—the wide boards running sideways to the

Archaeological site plan of the *Pretoria*

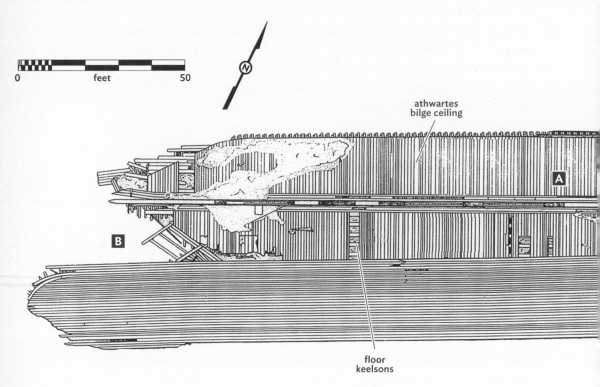

0   feet   50

athwartes
bilge ceiling

A

B

floor
keelsons

A. The open bilge of the *Pretoria* runs the length of a foot-ball field—a long swim for archaeologists when a strong current is running. TAMARA THOMSEN/WHI IMAGE ID 140412

B. Fragments of the shipwrecked *Pretoria* intermingle with the remaining ore. TAMARA THOMSEN/WHI IMAGE ID 140415

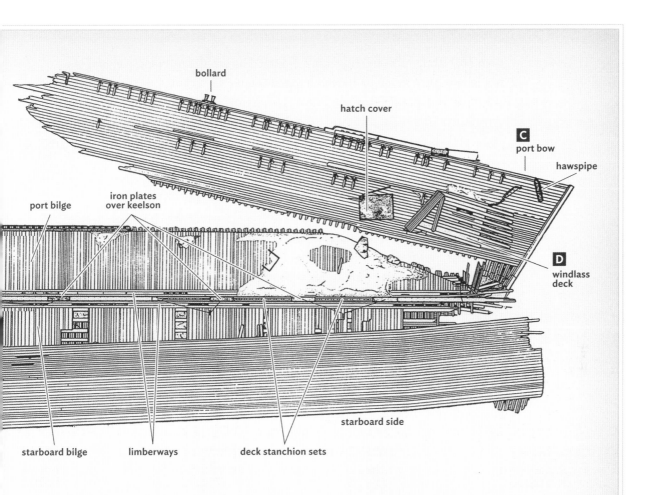

bollard

hatch cover

C
port bow

hawspipe

port bilge

iron plates
over keelson

D
windlass
deck

starboard side

starboard bilge   limberways   deck stanchion sets

C. The port bow section of the *Pretoria* collapsed outward, exposing ceiling plank and hawsepipe. TAMARA THOMSEN/ WHI IMAGE ID 140413

D. The *Pretoria* was slowly torn apart by the storm while anchored. Lying upside down on top of the bilge near its bow, the heavy anchor windlass remains attached to deck timbers. In the foreground are sections of the anchor chain. TAMARA THOMSEN, WHI IMAGE ID 141643

turn of the bilge toward the vessel center. The wrecking event, ice, and the occasional salvage diver have damaged the wreck to some degree, but the powerful oak planks, timbers, and steel straps and bands retain extreme durability. This makes the site capable of sustaining many decades or more of careful recreational diving.

Long and boxy with a single heavy keelson sandwiched between two 18-inch-deep by 1-inch-thick steel plates and a series of 12-inch by 12-inch floor keelsons, the bottom of the *Pretoria* has nearly the same architecture as the *Appomattox*, although the *Pretoria*'s keelson plates seem to be 1 inch deeper than the steamer's. There are, however, some discrepancies between data about the *Pretoria* published at the time of its launching and the archaeological findings. According to an account published in the British journal *Marine Engineering*, the *Pretoria*'s frames were 16 inches deep (molded) at the keelson, 14 at the turn of the bilge, and 7 inches at the top. The archaeologists recorded the first two of these measurements at 18 and 15 inches. The third measurement, 7 inches, was identical with the published data. The discrepancies in the first two measurements could be an error, or it could be that Davidson used heavier materials than what he reported. A systematic comparison between archaeological and historical records of the Davidson ships would make for an interesting future technical study.

On a fully intact Davidson vessel, it would be impossible to fully examine his use of steel bands and straps. On the *Pretoria*, the breaking away of the vessel's sides have exposed the 5-inch-wide by ½-inch-thick steel cross-brace strapping that tied together all of the ship's frames and extended well below the turn of the bilge into the bottom of the ship. Fortunately for the site's resilience but unfortunately for archaeologists, the well-fastened exterior planking and interior ceiling planking remain in place, hiding potentially unknown design elements. Especially impressive is the massive stem, a huge oak timber still connected to a series of angled cant frames. Among the other interesting features is the large steam windlass that rests upside down below an inverted section of bow decking.

The *Pretoria* is the least accessible of Wisconsin's Davidson shipwrecks. The location is fully exposed in Lake Superior and well off the mainland. On a clear day, one can see the Outer Island Lighthouse where the desperate crew of the *Pretoria* sought shelter. In 1994, the *Pretoria* was listed on the National Register of Historic Places.

Epilogue

THE END OF THE WOODEN AGE

It was not technological obsolescence that finally closed the era of wooden shipbuilding on the Great Lakes; it was simply the lack of building material. As long as inexpensive, high-quality oak was available, James Davidson could produce efficient moneymaking bulk freighters. The construction of the giant schooner barges *Chieftain* in October 1902 and *Montezuma* in July 1903 finally exhausted Davidson's readily available timber stocks.

On the Atlantic coast, the rapidly growing demand for coal to provide energy for industry and home heating and the continued availability of inexpensive timber sparked the largest shipbuilding boom in Maine history, with shipyards turning out massive four-, five-, and eventually six-masted wooden schooners. The New England coal schooners grew as dramatically and along nearly the same timeline as Great Lakes ships, and by 1890, the new vessels frequently exceeded 250 feet in length. They continued to grow during the first decade of the twentieth century, reaching the maximum length of 329.5 feet with William F. Palmer's six-masted schooner the *Wyoming*, launched in 1909.

Unlike the Great Lakes vessels built by Davidson and his competitors, the new Atlantic coast coal schooners showed little innovation in shipbuilding. Their New England builders, who retained access to inexpensive ship timber, simply enlarged the traditional wooden sailing ship, adding mass to the hulls, more masts, and donkey boilers and steam winches to save on labor. Even on that score, it was a Great Lakes shipyard that built the *David Dows*, the first schooner with five masts in the United States. Although some New England builders installed iron cross-bracing and belts similar to those used on the Great Lakes ships, when they did so, it was with substantially less iron.[1]

The schooner-barge *Chieftain* under construction at the Davidson yard in 1902.
C. PATRICK LABADIE COLLECTION / THUNDER BAY NATIONAL MARINE SANCTUARY, ALPENA, MI

In an unfair twist of history, by harkening back to the golden age of American sailing ships, the giant New England coal schooners immediately became and have remained romantic objects, inspiring generations of writers, scholars, and artists. By contrast, Davidson's giant schooners are the products of a progressive, forward-looking shipbuilder and independent industrial capitalist. Almost anti-romantic in their utilitarian design, they have, until recently, received little attention outside of the Great Lakes region.

The historical anonymity of James Davidson's giant schooners and his largest steamers illustrate marginalizing of the Great Lakes in American maritime history. Once clearly recognized as a nationally and even internationally important and highly creative maritime region with a distinct Atlantic frontier cultural heritage, the Great Lakes were demoted to inland waterway with a unique but purely regional significance and maritime history, cultural identity, and material culture. This transformation went hand in hand with, but was not the same as, the passing of the wooden age on the Great Lakes. Both transformations were almost but not fully complete by the time of the *Pretoria*'s loss in 1905. Mariners such as James Davidson who became thoroughgoing seamen through service on both salt water and fresh water were becoming scarce, and the cultural values, practices, and distinctive skills associated with traditional Atlantic merchant service slowly diluted. In a telling 1905 interview, sixty-two-year-old James Davidson responded wistfully about the decision of his youngest son, Harold Davidson, to enroll at a Chicago nautical school:

> Over there he will learn something we used to learn when I learned the business on the ocean. He will learn about the compass, its variations and adjustment and other things which the sailor on the lakes knows little about. . . . Although he may not follow sailing as a business a man who knows the business has the ordinary man beaten all around. The things a sailor knows can be applied to almost any trade or profession and he will never be sorry he knows what a sailor must know.[2]

The cultural divorce between the Great Lakes happened quickly, and the memories of the long relationship soon became obscured. In his famous 1884 report, federal investigator Henry Hall sang the praises of wooden ships and the shipyards of the Saginaw River as some of the finest in the United States. In 1889, William Wallace Bates, the builder of the schooner *Challenge* at Manitowoc and the designer of the *Magic* for the Ferry family in Grand Haven in the 1850s, was appointed US Commissioner of Navigation—at the time the most influential federal maritime post in the country. At the conclusion of his term in office, Bates published a massive treatise on the contemporary American merchant marine, whose place in international commerce had been in decline for decades. In a telling chapter on ship design, Bates counseled:

> Managers of American ocean lines of transportation have no need to look abroad for examples in types of vessels for economical transport. The best of these can be found in their own country, and on fresh water at that. We may declare with pride, that the shipbuilding, the navigation, and the management of transportation on the Lakes may challenge the world for quality.[3]

The view at the beginning of the twentieth century starts to become different. In 1902, Charles Scribner's Sons published journalist Winthrop L.

Marvin's *The American Merchant Marine: Its History and Romance from 1602 to 1902*. Marvin's book, like Bates's before it, was politically motivated and argued forcefully for a restored merchant marine. A journalist and close friend of Theodore Roosevelt, Marvin was a romantic whose desire to follow the footsteps of his old New England seafaring family had been crushed by the American merchant marine's precipitous decline. With the publication of his book, Marvin became a leading voice in merchant marine matters, and he was appointed secretary of the US Merchant Marine Commission created by Congress in 1904. His opinions carried weight in US maritime circles for decades and also influenced the writings of later maritime historians, who have tended to overlook the Great Lakes in American maritime history.

While describing the modernity and efficiency of the Great Lakes merchant marine and expressing admiration for "the skill and liberality of fresh water steam construction" and lake sailors, the "tall and massive Northern Vikings, clear-eyed, tanned with sun and sleet," Marvin dismissed the region's older wooden schooners and newer schooner barges as lacking "staunchness" and aesthetic merit: "The lake schooner of wood does not compare favorably in model with her new, powerful, and symmetrical sisters of the Atlantic and Pacific seaboard. She has too many suggestions in her straight up-and-down sides and blunt ends of the river scow or the canal-boat."[4] Marvin based these sweeping statements on the small "Grand Haven rig" schooners—an obsolete class of adapted vessels that, while undeniably efficient, had never been on the cutting edge of lake shipbuilding.[5] Carrying this description forward, in his otherwise excellent 1948 study of Atlantic coal schooners, W. J. Lewis Parker, citing Marvin as his source, dismissed all Great Lakes schooners as "greatly inferior in model and finish" to his primarily Maine-built subjects.[6] As a romantic, Marvin was enthralled by the Atlantic coasting schooners of the time:

> The modern coasting vessel of the United States is not a scow or a hulk, ill rigged, ill manned, shaky, and leaky, crawling from port to port when the wind serves and the sea is quiet. It is a thorough seagoing schooner, of two masts, three, four, five, six, and soon to be seven, able to look a gale right in the eye and to take whatever comes without flinching. American coasters of all types and sizes are loftily sparred beyond the cautious European practice, and, as a rule, are very deeply laden. Yet their models are so wholesome, and they are so resolutely handled by their few but good officers and men, that their work is done with no very high rate of disaster.[7]

It is with reference to these storied Atlantic coal schooners that David-son's *Pretoria* and the other great schooner barges he built between 1896 and 1902 are best judged. Because they could be built with a much greater depth, the largest Atlantic coal schooners could carry more cargo than the largest Davidson schooner barges, but a comparison of the 1900 schooner barge *Pretoria* alongside William F. Palmer's 1909 schooner *Wyoming*, the last and largest wooden six-master built, suggests some important differences, as shown in the following table.[8]

	Pretoria	*Wyoming*
Length	338.4	329.5
Width	44.0	50.1
Depth	23.0	30.4
Gross tons	2,790	3,730
Net tons	2,715	3,036
Capacity	5,100 tons	6,000 tons
Capacity per gross ton	1.83 tons	1.61 tons

The *Wyoming* was a third larger than the *Pretoria* in gross tons but carried only 18 percent more cargo. Relying upon traditional wooden shipbuilding methods, the builders of the coal schooners opted for structural mass, providing longitudinal support through hard pine keelsons built up six or seven feet, assisted by equally large sister keelsons, and ceiling planking often more than a foot thick. Hull strength also depended on many internal stanchions and required small deck hatches; this made loading and unloading the cargo particularly laborious and time consuming.[9]

James Davidson, who had abandoned traditional wooden hull designs in the 1870s, defied the 300-foot length barrier by carefully integrating steel into the hull structure. This practice added great strength and enhanced cargo capacity. In October 1900, a British trade journal, *Marine Engineering*, printed a detailed description of the *Pretoria*, including its steel reinforcements:

On the outside of the frames—at the head of the frames—there runs a steel cord that is 14 in. by 7-8 in. There are also steel arches running along the steel cords amidships to the fore foot forward and under the stern aft. In addition to these the *Pretoria* is diagonally strapped with steel. The diagonals are 5 in by 1-2 in. These diagonals straps are riveted to the cord, to the arch and at all of the crossings, and go under the turn of the bilge about one-third of the way under the bottom. The cords, arches and diagonal straps are all let into the frames flush, and then the ship is planked so that all of the strapping is concealed.[10]

In addition, as with the steamer *Appomattox*, Davidson added longitudinal strength with steel plates 18 inches wide by 1 inch thick running the length of *Pretoria's* main keelson. The *Wyoming* was also metal strapped, but with lighter bands of iron materially inferior to those long common in Great Lakes shipbuilding.[11] The *Pretoria's* deck arrangements with eleven wide, evenly spaced hatches, an idea pioneered thirty years earlier with the *R. J. Hackett*, and an open hold with its virtually flat bilge made loading and unloading far faster and more economical than with the *Wyoming* and other great coal schooners.

Despite their size and Marvin's statements to the contrary, the big coal schooners proved tender ships, exhibiting a "snake-like quality" in rough weather and leaking badly. In 1910, John J. Wardell, builder of the 1902 six-masted schooner *George W. Wells*, admitted that six-masters had proven impractical because they were too long for wooden construction. These problems worsened with age, and nearly all of the eleven six-masted schooners built in New England met violent ends.[12]

Accidents on the Great Lakes remained unavoidable at the beginning of the twentieth century, and Davidson's giant schooner barges had their share of groundings, collisions, and catastrophic storms, with three ultimately

The wooden six-masted coal schooner *Wyoming*, built in Maine in 1909, had greater tonnage and capacity than the largest Davidson schooner barges but was not as long, nor as efficient or durable. MAINE MARITIME MUSEUM

lost. As a class, however, they proved durable, resilient, and seaworthy, with the seven largest craft, all more than 300 feet in length, enjoying an average service life of well over twenty years. Only *Pretoria*, sunk at age five after losing its hydraulic steering in a devastating Lake Superior storm, had a short career. Most impressively, three out of four of the final schooner barges produced by Davidson, including the *Montezuma* and the *Chieftain*, the longest wooden schooner hulls ever built in the United States, survived three decades of hard service before abandonment in the Saginaw River.[13]

As examples of wooden ship architecture, the Davidson schooner barges far exceeded the Atlantic schooners in strength and operational efficiency. However, as parts of an industrial transportation system that relied on steam and tight schedules, the boxy vessels lacked the romance and visual appeal of the ocean craft. However, the Davidson schooner barges—indeed, all of the wooden vessels built on the Great Lakes during the nineteenth century—were products of the same Atlantic maritime culture. The Great Lakes maritime frontier provided distinctive economic opportunities, unique geographic influences, and, just perhaps, a few degrees of freedom from the cultural conventions and old rules that constrained the creativity of shipbuilders in the older Atlantic states. At least, that is the story told by Wisconsin shipwrecks.

Acknowledgments

This book could not have been written without the remarkable public investment in preserving, studying, and interpreting for the public Wisconsin's more than six hundred historic Great Lakes shipwrecks. The decision made during an early-morning committee session of the state legislature in the late 1980s to establish a state underwater archaeology program has continued to pay remarkable dividends to the people of Wisconsin. Thousands of children have studied Wisconsin's maritime history and learned the critical-thinking and detective skills of the underwater archaeologist through educational materials and programs created through the efforts of the Wisconsin Historical Society and its many public and private partners across the state. The installation and maintenance of mooring buoys and integrated maritime heritage trails have made visiting shipwrecks safer for people and less damaging for the wrecks. Such efforts have contributed to heritage tourism in coastal destinations, attracting and improving the experience of visitors of all ages and backgrounds. In terms of preservation, Wisconsin leads all states in the number of shipwrecks listed on the National Register of Historic Places—nearly sixty at the time of this book's completion. All this work required tens of thousands of hours put in by countless people, many of them individual volunteers or members of preservation organizations such as the Wisconsin Underwater Archaeological Society (WUAA) and the Great Lakes Shipwreck Preservation Society (GLSPS).

While funding for its underwater archaeology and maritime heritage program has waxed and waned with economic and political circumstances, the Wisconsin Historical Society has benefited from continuing support from other public agencies. From the earliest days of the State Underwater Archaeology program, the University of Wisconsin Sea Grant Institute has provided continuing support for research funding and equipment and has been a crucial partner in a wide area of outreach and public education efforts. All the underwater archaeology projects that directly contributed to the shipwreck-inspired histories included in this book received Sea Grant support. Among the many other partnering and funding agencies are the Wisconsin Department of Transportation, the Wisconsin Coastal Zone Management Program, the University of Wisconsin–Milwaukee Great Lakes Water Institute, the Wisconsin Department of Natural Resources, the National Park Service, and the National Oceanographic and Atmospheric Administration Office of National Marine Sanctuaries.

The chapters in this book all began taking shape as individual projects to document historic shipwrecks in Wisconsin. All archaeological data and substantial elements of the original historical research were generated through the team and individual efforts of dozens of people. Crucial to this book is my old partner and Wisconsin's first state underwater archaeologist, David J. Cooper, who hired me as his assistant in 1990. Now with the National Park Service at the Apostle Islands National Lakeshore, David established an entrepreneurial and public-oriented professional culture for the State Underwater Archaeology Program that remains in place today. Jeff Gray, Russ and Cathy Green, Keith Meverden, Tamara Thomsen, and Caitlin Zant have carried on Coop's legacy, each leaving his or her own stamp on Wisconsin underwater archaeology. All have contributed directly to the completion of this book. Former and current state archaeologists Robert Birmingham and John Broihahn have made exceptional efforts to promote maritime historic preservation and underwater archaeology in Wisconsin and provided me with many professional opportunities that underpin this book.

Most of the contemporary underwater photographs in this book are the work of Tamara Thomsen, Preservation Specialist with the Wisconsin Historical Society Maritime Preservation and Archaeology Program. She is an unparalleled shipwreck photographer, and her contributions communicate the wonder and meaning of Wisconsin shipwrecks more clearly than any words. She also read and provided valuable comments on the first draft of the manuscript.

During the financially strained times of 1994 when the work on the steamer *Niagara* was being conducted, individual members of the Wisconsin Underwater Archaeological Association provided funds for a research boat as well as their time and skills to help complete the project; this is just one example of the extraordinary public investment in preserving the state's shipwrecks. Generations of graduate students and the faculty with the Program in Maritime History and Nautical Archaeology at East Carolina University, currently directed by Green Bay, Wisconsin, native Bradley Rogers, have contributed to dozens of shipwreck projects in Wisconsin over the decades, including work on the schooner *Lucerne* and the James Davidson vessels.

The availability of primary printed sources to study nineteenth-century Great Lakes history has expanded exponentially since the early days of the underwater archaeology program. All of us in the Great Lakes history community are indebted to historian C. Patrick Labadie, formerly of the Canal Park Marine Museum and the Thunder Bay National Marine Sanctuary. The most knowledgeable living authority on Great Lakes ships and shipping history, Pat was continuously generous with his time, knowledge, and materials from his vast personal collection—which he donated to the public and is now housed at the George N. Fletcher Public Library in Alpena, Michigan. Key parts of his collection have been digitized and are available through the Fletcher Library website, http://greatlakeships.org/search. Pat's inspiration and influence on my work is beyond calculation; he provided foundational information for the early work on the James Davidson ships and the palace steamer *Niagara* and helped with the underwater documentation. He generously read drafts of key chapters, and I hope that the book comes close to the standards he has set for Great Lakes history over the past fifty years.

An early wizard at digitizing Great Lakes historical sources, Brendon Baillod provided a great deal of research material to the State Underwater Archaeology Program that was used in this book. Anyone researching Great Lakes maritime history today is indebted to librarian and historian Walter Lewis for creating and maintaining the extraordinary digital library Maritime History of the Great Lakes. Dan Johnson provided access to artifacts from the schooners *Lumberman* and *Kate Kelly* housed at the Lockwood Pioneer Diving Museum in Loves Park, Illinois, and valuable information on wreck sites and on the history of Captain Hartley Hatch.

This book has taken more than ten years from its conception to complete. Despite this inexcusable amount of time, the editors at the Wisconsin Historical Society Press have been patient and supportive. Special thanks go to Kathy Borkowski, Kate Thompson, and manuscript editors Barbara Walsh and Erika Wittekind. Working with a dyslexic and obstinate ex–king crab fisherman (even one with a PhD) is not easy, and I appreciate their efforts. Archivists at the Fletcher Library in Alpena, Michigan, the Wisconsin Historical Society, the National Archives Chicago Branch, the Milwaukee Public Library, and Clements and Bentley Libraries at the University of Michigan in Ann Arbor were all very generous and helpful in finding sources. Tamara Thomsen, Caitlin Zant, Victoria Kiefer, and Tom Villand have been generous with their assistance and access to the research files of the Wisconsin Historical Society Maritime Preservation and Archaeology Program.

Graduate students from the University of West Florida Department of History Abigail Melton, Ian Hamilton, and Jessie Craig all read and assisted in the completion of book chapters. Phil Hartmeyer, archaeologist with the Thunder Bay National Marine Sanctuary, also provided valuable research assistance. The latter stages of the book benefited from a generous faculty research stimulus grant from the University of West Florida Office of Research and Sponsored Programs.

Although this book was partially funded and published by the Wisconsin Historical Society, I am responsible for the overall historical interpretations offered in this book and for any errors. The core themes have evolved through a lifetime of experience working on the ocean and lake, nearly thirty years of research, and endless conversations with my Great Lakes colleagues and a wide array of maritime scholars, most significantly Rod Mather, David J. Cooper, Russ Green, Jeff Gray, and Patrick Labadie. Also influential have been the comments and enthusiasm of the dozens of scholars from around the United States who participated in my maritime frontier sessions in summer graduate programs and National Endowment for the Humanities seminars offered through the Munson Institute at Mystic Seaport Museum in Mystic Connecticut.

Finally, I must thank my family, especially my wife, historian of technology Karen Belmore, and my aunt Odeena Jensen Strange. Born on an island in Alaska in 1921, Odeena is a living example of the strength, adaptability, cosmopolitanism, and common sense of Scandinavian maritime people. She gave me no choice but to finish this book.

BIBLIOGRAPHY

Abbot, Jacob. "The Novelty Works, With Some Descriptions of the Machinery and the Processes Employed in the Construction of Marine Steam-Engine of the Largest Class." *Harper's New Monthly Magazine*, no. 2 (1851): 721–34.

"An Act to Limit the Liability of Ship-Owners, and for Other Purposes" (31st Cong., 2nd Sess.), 1851. loc.gov /law//help/statutes-at-large/31st-congress/session-2 /c31s2ch43.pdf.

Albion, Robert Greenhalgh. *The Rise of New York Port [1815–1860].* Princeton, NJ: Princeton University Press, 1938.

Albion, Robert Greenhalgh. *Square-Riggers on Schedule; the New York Sailing Packets to England, France, and the Cotton Ports.* Princeton, NJ: Princeton University Press, 1938.

Alden, Henry Mills. "The Old Shipbuilders of New York." *Harper's New Monthly Magazine*, November 1882: 223–41.

Alford, Harold D. "Shipbuilding Days in Old Oswego." *Inland Seas* 13, no. 2–4 (1957): 88–94, 219–25, 306–12.

Allen, Ned. *The Sebago Lakes Region: A Brief History.* Charleston, SC: The History Press, 2013.

American Board of Commissions for Foreign Missions. *Report of the American Board of Commissions for Foreign Missions.* Boston: Crocker & Brewster, 1827.

Ancestry.com. "Pennsylvania Biographical Sketches, 1868." Provo, UT: Ancestry.com Operations, 2000.

"*Australasia* National Register Nomination." Washington DC: US Department of the Interior National Park Service, 2013. www.nps.gov/nr/feature/places/pdfs /13000466.pdf.

Bajema, Carl Jay. "Blendon Landing: A Middle Nineteenth Century Logging Railroad, Sawmill, and Shipyard in West Michigan." *The Michigan Archaeologist* 43, no. 2–3 (1997): 103–14.

Bajema, Carl Jay. "The First Logging Railroads in the Great Lakes Region." *Forest & Conservation History* 35, no. 2 (1991): 76–83.

Baker, Catherine. *Shipbuilding on the Saginaw.* Bay City, MI: Museum of the Great Lakes, 1974.

Barkhausen, Henry. *Focusing on the Centerboard.* Manitowoc, WI: Association for Great Lakes Maritime History, 1990.

Barton, James L. *A Brief Sketch of the Commerce of the Great Northern and Western Lakes For a Series of Years to Which Is Added, an Account of the Business Done through Buffalo on the Erie Canal for the Years 1845 and 1846.* Buffalo, NY: Jewett and Thomas, 1847.

Bascom, John H., and Peter J. Van der Linden, eds. *Great Lakes Ships We Remember.* Cleveland, OH: Freshwater Press, 1979.

Baskerville, Peter A. "Donald Bethune's Steamboat Business: A Study of Upper Canadian Commercial and Financial Practice." *Ontario History* 67 (1975): 135–49.

Bates, William W. *American Marine: The Shipping Question in History and Politics.* Boston: Houghton Mifflin, 1892.

Bates, William W. "Inland Navigation." *Monthly Nautical Magazine and Quarterly Review* 1 (1854): 9–16.

Bates, William W. "Marine Insurance and Its Influence on Shipbuilding." *U.S. Nautical Magazine and Naval Journal* 5, 1 (1856): 1–8.

Bates, William W. "Mould-Loft Tables of the Schooner Magic." *Monthly Nautical Magazine and Quarterly Review* 2 (1855): 498–501.

Bates, William W. "The Tables of the Schooner Challenge." *Monthly Nautical Magazine and Quarterly Review* 2 (1855): 106–11.

Bellamy, John S. *The Last Days of Cleveland and More True Tales of Crime and Disaster from Cleveland's Past.* Cleveland, OH: Gray and Company, 2010.

Bird, Isabella Lucy. *The Englishwoman in America.* London: John Murray, 1856.

Blois, John T. *Gazetteer of the State of Michigan, in Three Parts.* Detroit: Sydney L. Rood, 1839.

The Blue Book of American Shipping 1900. Cleveland, OH: Marine Review Publishing, 1900.

Board of Lake Underwriters. *Classification of Lake Vessels and Barges*. Buffalo, NY: Warren, Johnson, and Co.: 1871.

"The Board of Lake Underwriters." *U.S. Nautical Magazine and Quarterly Review* 6, no. 3 (1857): 190–94.

Bowlus, W. Bruce. *Iron Ore Transport on the Great Lakes: The Development of a Delivery System to Feed American Industry*. Jefferson, NC: McFarland, 2010.

Britten, Emma Hardinge. *Modern American Spiritualism; a Twenty Years' Record of the Communion between Earth and the World of Spirits*. New Hyde Park, NY: University Books, 1899.

Buckingham, J. S. *The Eastern and Western States of America*. Vol. 3. London: Fisher, Son, & Co, 1843.

Buffalo City Directory. Buffalo, NY: Thomas, Howard, and Johnson, 1866.

"Enrolled at Port of Buffalo, Number 10 of 1829," September 2, 1829. Transcribed at Maritime History of the Great Lakes (MHGL), www.maritimehistory ofthegreatlakes.ca/Enrolment/Enrolment.asp ?EventID=160.

Cangany, Catherine. *Frontier Seaport: Detroit's Transformation into an Atlantic Entrepôt*. Chicago: University of Chicago Press, 2014.

Cantelas, Frank J. "An Archaeological Investigation of the Steamboat Maple Leaf." Master's thesis, East Carolina University, 1995.

Chernow, Ron. *Titan: The Life of John D. Rockefeller, Sr.* New York: Vintage Books, 2004.

Church, William C. *The Life of John Ericsson*, Volume I. New York: Charles Scribner's Sons, 1906.

Circular and Catalogue of Bryant and Stratton Mercantile Colleges. New York: Office of the American Merchant, 1859.

Classification of Lake Vessels and Barges: Adopted by a Board of Marine Inspectors. Buffalo, NY: Warren, Johnson & Co, 1871. http://images.maritimehistory ofthegreatlakes.ca/65698/data.

"Clipper *Panther*—Famous Ships of the British Columbia Coast." Shady Isle Pirate Society, 2004. http://bbprivateer.ca/panther.

Coombe, Philip W. "James P. Allaire: Marine Engine Builder." *Steamboat Bill of Facts* 43 (1986): 263–71.

Coombe, Philip W. "The Life and Times of James P. Allaire: Early Founder and Steam Engine Builder." PhD diss., New York University, 1991.

Cooper, David J. *Survey of Submerged Cultural Resources in Northern Door County: 1988 Field Season Report*. Madison: Office of the State Archaeologist, Historic Preservation Division, State Historical Society of Wisconsin, 1989.

Cooper, David J., et al. *By Fire, Storm, and Ice: Underwater Archeological Investigations in the Apostle Island*. Madison: State Historical Society of Wisconsin, 1991.

Cooper, David J., and John O. Jensen. *Davidson's Goliaths: Underwater Archaeological Investigations of the Steamer Frank O'Connor and the Schooner-Barge Pretoria*. Madison: State Historical Society of Wisconsin, State Underwater Archaeology Program, 1995.

Crisman, Kevin J. *Coffins of the Brave: Lake Shipwrecks of the War of 1812*. College Station: Texas A&M University Press, 2014.

Cronon, William. *Nature's Metropolis: Chicago and the Great West*. New York: W. W. Norton & Company, 1992.

Dana, Jr., Richard Henry. *Two Years Before the Mast: A Personal Narrative of Life at Sea*. New York: New American Library, 1985.

Dickson, Kenneth R. "The Longest Schooner (1881) in the World." *Inland Seas* 42 no. 1 (1986): 2–6.

"Disasters on the Lakes." *U.S. Nautical Magazine and Naval Journal* 5 no. 3 (1856): 3.

Dreyful, Benjamin. "The City Transformed: Railroads and Their Influence on the Growth of Chicago in the 1850s." Harvard University, 1995. www.hcs.harvard .edu/~dreyfus/history.html.

Errols, Robert. "Thurlow Weed's Report." In *Chicago Rivers and Harbors Commission: An Account of Its Origin and Proceedings*. Chicago: Fergus Printing, 1882.

Findlay, James. F. *Dwight L. Moody, American Evangelist, 1837–1899*. Eugene, OR: Wipf and Stock: Eugene, 2007.

Games, Alison. "Introductions, Definitions, and Historiography: What Is Atlantic History?" *Organization of*

American Historians Magazine of History 18, no. 3 (2004): 3–7.

"General Charles M. Reed." *Magazine of Western History* V (January 1887): 400–10.

Gould, Richard A., *Shipwreck Anthropology*. Albuquerque, NM: School for Advanced Research Press, 1983.

Grant, James. *Recent British Battles on Land and Sea*. London: Cassell & Company, 1884.

"Great Lakes Vessels Index." Historical Collections of the Great Lakes, Bowling Green State University, http://greatlakes.bgsu.edu/vessel.

Griffiths, John W. *Treatise on Marine and Naval Architecture, Or, Theory and Practice Blended in Ship Building*. New York: D. Appleton, 1852.

Hales, David A. and Sandra Dawn Brimhall. "William and Jeannette Ferry: Presbyterian Pillars in Mormon Utah." *Utah Historical Quarterly* 79, no. 2: 122–143.

Hall, Henry. *Report on the Ship-Building Industry of the United States*. Washington, DC: US Government Printing Office, 1884.

Hall, Henry. "Ship Building in the United States." Notebook on file in the Penobscot Marine Museum, Searsport, Maine, 1881.

Hall, James. *The West: Its Commerce and Navigation*. Cincinnati, OH: W. W. Derby, 1848.

Hardinge, Emma. *Modern American Spiritualism: A Twenty-Year Record of Communion between Earth and the World of Spirits*. New York: Emma Hardridge, 1870.

Haswell, Charles Haynes. *Engineers' and Mechanics' Pocket-Book*. New York: Harper & Brothers, 1882.

Havighurst, Walter. *The Long Ships Passing: The Story of the Great Lakes*. New York: Macmillan, 1975.

Heath, May Francis. *Early Memories of Saugatuck, Michigan: 1830–1930*. Grand Rapids, MI: Wm. B. Eerdmans, 1930.

Herdendorf, Charles E., and Sandra E. Schuessler. *The First Half Century of Merchant Steam Vessels on the Lakes*. Put-in-Bay: Ohio State University Center for Lake Erie Research, 1993.

Hilton, George Woodman. *The Great Lakes Car Ferries*. Berkley, CA: Howell-North, 1962.

Hilton, George Woodman. *Lake Michigan Passenger Steamers*. Stanford, CA: Stanford University Press, 2002.

Historical Society of Pennsylvania, *Encyclopedia of Contemporary Biography of Pennsylvania*, vol. II. Bethlehem, PA: Historical Society of Pennsylvania, 1868. Retrieved from Ancestry.com.

History of Bay County, Michigan. Chicago: H. R. Page, 1883.

The History, Commercial Advantages, and Future Prospects of Bay City. Bay City: Henry S. Dow, 1875.

Hogan, William Thomas. *Economic History of the Iron and Steel Industry in the United States*. Lexington, MA: Heath, 1971.

Holzheuter, John O. "Wisconsin's Flag." *Wisconsin Magazine of History*, Winter 1979–1980: 90–121.

Hotchkiss, George Woodward. *Industrial Chicago; the Lumber Interests*. Chicago: Goodspeed, 1894. http://hdl.handle.net/2027/mdp.39015071163896.

Howe, Octavius, and Frederick Mathews. *American Clipper Ships 1833-1858*. Mineola, NY: Dover, 1986.

Hunter, Louis C. *Steamboats on the Western Rivers: An Economic and Technological History*. Cambridge, MA: Harvard University Press, 1949.

James, Thomas Horton. *Rambles in the United States and Canada during the Year 1845, with a Short Account of Oregon*. London: J. Ollivier, 1846.

Jensen, John O. "The Great Lakes." In Vol. 2, *The Oxford Encyclopedia of Maritime History*, edited by John B. Hattendorf, 101–11. Oxford University Press, 2007.

Jensen, John O. 1999. "The History and Archeology of the Great Lakes Steamboat Niagara." *The Wisconsin Magazine of History* 82, no. 3 (spring 1999): 198–230.

Jensen, John O. "Oak Trees and Balance Sheets: James Davidson, Great Lakes Shipbuilder and Entrepreneur." *American Neptune* 54, no. 2 (spring 1994): 99–114.

"Joseph Holmes Obituary." *Maryland Historical Magazine* 19 (1919): 299–300.

Karamanski, Theodore. *Schooner Passage: Sailing Ships and the Lake Michigan Frontier*. Detroit, MI: Wayne State University Press, 2000.

Kilar, Jeremy W. *Michigan's Lumbertowns: Lumbermen and Laborers in Saginaw, Bay City, and Muskegon, 1870–1905*. Vol. 1 and 2. Detroit, MI: Wayne State University Press, 1990.

Labadie, C. Patrick. *Submerged Cultural Resources Study: Pictured Rocks National Lakeshore*. Southwest Cultural Resources Center Professional Papers 22. Santa Fe, NM: National Park Service Southwest Cultural Resources Unit, 1989.

Labaree, Benjamin, et al. *America and the Sea: A Maritime History*. Mystic, CT: Mystic Seaport Museum, 1998.

Lake Vessel Register, System of Classification. Buffalo, NY: Board of Lake Underwriters, 1874.

Lane, Carl D. *American Paddle Steamboats*. New York: Coward-McCann, 1943.

Lenihan, Daniel J. *Shipwrecks of Isle Royale National Park: The Archeological Survey*. Duluth, MN: Lake Superior Port Cities, 1994.

Lewis, Walter. "The Frontenac: A Reappraisal." *Fresh Water* 2, no. 1 (1987): 28–39. www.maritimehistoryof thegreatlakes.ca/Documents/Frontenac.

Lillie, Leo C. *Historic Grand Haven and Ottawa County*. Grand Haven, MI: Leo C. Lillie, 1931.

MacKinnon, Lauchlan Bellingham. *Atlantic and Transatlantic: Sketches Afloat and Ashore*. New York: Harper Brothers, 1852.

Maddox, Melvin. *The Atlantic Crossing*. New York: Time-Life Books, 1982.

Mansfield, John B. *The History of the Great Lakes*. Chicago: J. H. Beers, 1899.

Marestier, Jean Baptiste. *Memoir on Steamboats of the United States of America*. Hartford: Connecticut Printers, 1957.

Marjoribanks, Alexander. *Travels in South and North America*. London: Simpkin, Marshall, and Company, 1854.

Marley, John F. *The Life of Reverend Thomas A. Morris, D.D.* Cincinnati, OH: Hitchcock and Walden, 1875.

Martin, Jay C. "The Grand Haven Rig: A Great Lakes Phenomenon." *American Neptune* 51, no. 3 (1991): 195–201.

Marvin, Winthrop Lippitt. *The American Merchant Marine; Its History and Romance from 1620 to 1902*. New York: C. Scribner's, 1902.

Marx, Dede, and Mathew Lawrence. "National Register of Historic Places Registration Form: Paul Palmer—Shipwreck and Remains." n.d. On File, National Park Service, Keeper of the National Register's office.

McCarty, John Myron. "Economic Aspects in the Evolution of the Great Lakes Freighter." PhD diss., University of South Carolina, 1971.

McDonald, William A. "Eyewitness Reports of the Burning of the Niagara." *Inland Seas* 12 (1956): 195–202.

Meverden, Keith N., and Tamara L Thomsen. *Myths and Mysteries: Underwater Archaeological Investigation of the Lumber Schooner Rouse Simmons, Christmas Tree Ship*. State Archaeology and Maritime Preservation Program Technical Report Series 08–01. Madison: State Historical Society of Wisconsin Press, 2008.

Meverden, Keith N, and Tamara L Thomsen. *Wisconsin's Cross-Planked Mosquito Fleet: Underwater Archaeological Investigations of the Scow Schooners Iris, Ocean Wave, and Tennie and Laura*. State Archaeology and Maritime Preservation Program Technical Report Series 06-001. Madison: Wisconsin Historical Society, 2006.

Meverden, Keith N., Tamara L. Thomsen, and John O. Jensen. *Wheat Chaff and Coal Dust: Underwater Archaeological Investigations of the Grain Schooners Daniel Lyons and Kate Kelly*. State Archaeology and Maritime Preservation Technical Report Series 06-02. Madison: Wisconsin Historical Society, 2006.

Morris, Paul C. *American Sailing Coasters of the North Atlantic*. Chardon, OH: Bloch and Osborn Publishing Co, 1973.

Morris, Thomas A. *Miscellany: Consisting of Essays, Biographical Sketches and Notes of Travel*. Cincinnati, OH: L. Swormstedt & A. Poe, 1852.

Morrison, John H. 1903. *History of American Steam Navigation*. New York: W. F. Sametz, 1903.

Morrison, John H. *A History of New York Shipyards*. New York: W. F. Sametz, 1909.

Mott, Edward Harold. *Between the Ocean and the Lakes*. New York: Collins, 1899.

Murphy, Larry. "Shipwrecks as Database for Human Behavioral Studies." Chap. 4 in *Shipwreck Anthropology*, edited by Richard Gould. Albuquerque: University of New Mexico, 1993.

Nelson's Biographical Dictionary and Historical Reference Book of Erie County, Pennsylvania. Erie, PA: S. B. Nelson, 1896.

Odle, Thomas D. "Great Lakes History and the St. Clair Flats." *Detroit Historical Society Bulletin* 15 (1959).

Oldham, Joseph R. "Shipbuilding and Transportation on the Great American Lakes." *Cassier's Magazine* 12 (1896): 499–512.

Olsen, Valerie. *The Lady Elgin: A Report on the 1992 Reconnaissance Survey.* Chicago: Underwater Archaeological Society of Chicago, 1993.

Palmer, Richard F. "The First Steamboat on the Great Lakes." *Inland Seas* 44, no. 1 (1968): 7–20.

Parker, W. J. Lewis. *The Great Coal Schooners of New England, 1870–1909.* Mystic, CN: The Marine Historical Association: 1948.

Parry, J. H. *Discovery of the Sea.* Oakland: University of California Press, 1974.

Paskoff, Paul F. *Troubled Waters: Steamboat Disasters, River Improvements, and American Public Policy, 1821–1860.* Baton Rouge: Louisiana State University Press, 2007.

Plumb, Ralph G. "The 1857 Depression on the Lakes." *Inland Seas* 10, no. 4 (1954): 290–91.

Portrait and Biographical Record of Muskegon and Ottawa Counties, Michigan. Chicago: Biographical Publishing Co, 1893.

Quinney, Valerie. *Bryant College: The First 125 Years.* Smithfield, RI: Bryant College, 1988.

Rae, James D. "Great Lakes Commodity Trade, 1850–1900." PhD diss., Purdue University, 1967.

Railroad Convention. *Proceedings of the Rail-Road Convention, Assembled at Harrisburg, March 6, 1838.* Philadelphia: P. Hay, 1838.

Reeves, Pamela Wilson. "Navigation on Lake Erie." *Inland Seas* 15 (1960).

Richards, Thomas. "Letter to Louisa Richards," October 25, 1846. Janet Tuttle, personal collection.

Rodgers, Bradley A. 2003. *The Bones of a Bulk Carrier: The History and Archaeology of the Wooden Bulk Carrier/Stone Barge City of Glasgow.* East Carolina Program in Maritime History and Nautical Archaeology, Research Report 12. Greenville: East Carolina University Program in Maritime History and Nautical Archaeology, 2003.

Russell, Norman S. "On American River Steamers." In *Transactions of the Royal Institution of Naval Architects*, London: Royal Institution of Naval Architects, 1861.

"Samuel F. Hodge." *Magazine of Western History* 4, no. 5 (1886): 649–53.

Sanford, Laura G. *The History of Erie County, Pennsylvania: From Its First Settlement.* Erie, PA: Laura G. Sanford, 1862.

Schneider, C. F. *Michigan Section of the Climate and Crop Service of the Weather Bureau Report for November 1905.* Grand Rapids, MI: Weather Bureau Office, 1905.

Schoolcraft, Henry Rowe. *Personal Memoirs of a Residence of Thirty Years with the Indian Tribes on the American Frontiers; With Brief Notices of Passing Events, Facts, and Opinions, A.D. 1812 to A.D. 1842.* Philadelphia: Lippincott, 1851.

Seibert, Wilbur. *The Underground Railroad From Slavery to Freedom.* London: MacMillan, 1898.

Shipley, Robert, and Fred Addis. *Paddlewheelers.* St. Catharines, ON: Vanwell, 1991.

Smith, Alice E. *The History of Wisconsin Volume I: From Exploration to Statehood.* Madison: State Historical Society of Wisconsin Press, 1973.

Smith, Russell D. "The Northern Railway: Its Origins and Construction, 1834–1855." *Ontario History* 48: 24–36, 1956).

Spencer, Herbert, R. "The Reed Mansion." *Inland Seas* 20, no. 4: 313–23, 1964.

Stevenson, David. *Sketch of the Civil Engineering of North America.* London: John Weale, 1837.

Stone, Joel Lagrou. 2015. *Floating Palaces of the Great Lakes: A History of Passenger Steamships on the Inland Seas.* Ann Arbor: University of Michigan Press, 2015.

Swayze, David. 1991. "The Giant Wooden Barges of the Davidson Yard" *Inland Seas* 47, no. 2: 104–8.

Taylor, George R. *Transportation Revolution, 1815–1860.* London: Routledge, 1977.

Thiesen, William H. *Industrializing American Shipbuilding: The Transformation of Ship Design and Construction, 1820–1920.* Gainesville: University Press of Florida, 2006.

Thomas, Robert. *Register of the Ships of the Lakes and River St. Lawrence, 1864.* Buffalo, NY: Wheeler, Mathews, and Warren, 1864.

Thompson, Mark L. *Graveyard of the Lakes.* Detroit, MI: Wayne State University Press, 2000.

Thomsen, Tamara, et al. *With Waves in Their Wake: Underwater Archaeological Investigations from the 2013 Field Season.* State Archaeology and Maritime Preservation Technical Report Series 14-002. State Archaeology and Maritime Preservation Technical Report Series 13-001. Madison: Wisconsin Historical Society, 2014.

Thomsen, Tamara, et al. *Wisconsin Coal Haulers Underwater Archaeological Investigations from the 2012 Field Season.* State Archaeology and Maritime Preservation Technical Report Series 13-001. Madison: Wisconsin Historical Society, 2013.

Thomsen, Tamara L., C. N. Zant, and T. Kiefer. *Great Lakes Sailing Canallers and Other Underwater Archaeological Investigations from the 2016 Field Season.* State Archaeology and Maritime Preservation Technical Report Series 13-001. Madison: Wisconsin Historical Society, 2017.

Thomson, Ross. *Structures of Change in the Mechanical Age: Technological Innovation in the United States, 1790–1865.* Baltimore: Johns Hopkins University Press, 2009.

Tunell, George Gerrard. "Transportation on the Great Lakes of North America." PhD diss., New York University, 1890. Published as H. doc. 277. 55th Cong. 2nd Sess. Washington, DC: GPO, 1898.

United States. *Acts of Congress Relating to Steamboats, Collated with the Rolls at Washington.* Boston: Little, Brown, 1853.

United States Life-Saving Service. *Annual Report of the United State Life-Saving Service for the Fiscal Year Ending June 30, 1906.* Washington DC: Government Printing Office, 2007.

US Bureau of Navigation. *List of Merchant Vessels of the United States.* Washington DC: Government Printing Office, various years.

US Census Office, *Ottawa County, Michigan, Population Schedule: 1860.* Washington DC: Government Printing Office, 1860. Retrieved from Ancestry.com.

US Census Office, *Report on Manufacturing Industries in the United States: 1890. Part III, Selected Industries.* Washington DC: Government Printing Office, 1890. 550–55.

US Patent Office. *Specifications of Patents Issued from the United States Patent Office for the Week Ending April 30, 1872.* Washington DC: Government Printing Office, 1872.

Van der Linden, Peter J. *Great Lakes Ships We Remember.* Cleveland, OH: Freshwater Press, 1979.

Wade, Richard C. *The Urban Frontier: The Rise of Western Cities, 1790–1830.* Urbana: University of Illinois Press, 1996.

Walker, Augustus. "Early Days on the Lakes with the Account of the Cholera Visitation of 1832." *Publications of the Buffalo Historic Society* (January 1902): 287–318.

Widder, Keith R. *Battle for the Soul: Mètis Children Encounter Evangelical Protestants at Mackinaw Mission, 1823–1837.* Lansing: Michigan State University Press, 1999.

Whitaker, H. "The Influence of Model upon the Running Qualities of Steamboats." *U.S. Nautical Magazine and Quarterly Review* 6: 102–4, 1857.

Whitman, Benjamin. *Nelson's Biographical Dictionary and Historical Reference Book of Erie County.* Erie, PA: S. B. Nelson, 1896.

Williams, Mentor L. "The Chicago River and Harbors Convention, 1847." *Mississippi Valley Historical Review* 35 (1949).

Williams, Michael. *Americans and Their Forests: A Historical Geography.* Cambridge: Cambridge University Press, 1989.

———. "Products of the Forest: Mapping the Census of 1840." *Journal of Forest History* 24, no. 1: 4–23, 1980.

Williamson, Samuel H. "The Growth of the Great Lakes as a Major Transportation Resource, 1870–1911." *Research in Economic History: An Annual Compilation of Research* 2, 1977.

Wood, Donald. "A Morning with Valentine Fries, Shipbuilder." *Inland Seas* 41: 172–74, 1985.

Wolff, Julius R. "The Shipwrecks of Lake Superior, 1900–1909." *Inland Seas* 27: 174–84, 1971.

Wright, Richard J. *Freshwater Whales; a History of the American Ship Building Company and Its Predecessors.* Kent, OH: Kent State University Press, 1970.

Archival Sources Cited

Bentley Historical Library, University of Michigan, Ann Arbor

Ferry Family Papers (William Montague Ferry family): 1823–1904.

Gilchrist Family Papers: 1867–1945.

William L. Clements Library, University of Michigan, Ann Arbor

Litchfield-French Papers.

Milwaukee Public Library Great Lakes Marine Collection

Vessel Enrollments—*Kate Kelly, Lumberman.*

National Archives Chicago Branch

Record Group 21

US District Court, North District of Ohio, Eastern Division, Cleveland. Admiralty Records, Admiralty Case Files, 1855–1966. Case nos. 1515, 1527, 1531, 1532, 1533, 1534, 1535.

Penobscot Marine Museum, Searsport, Maine

Henry Hall's *Notebooks for Shipbuilding in the United States.*

Wisconsin Historical Society Archives

Map of the Counties of Ottawa & Muskegon and Part of Allegan, Michigan: From Official Records & Special Surveys. By I. M. Gross Under the Direction of Geil & Harley; Engraved by Worley & Bracher 1864. Digital copy available: http://cdm15932.contentdm.oclc.org /cdm/ref/collection/maps/id/15939.

Key Digital Libraries

Alpena County George N. Fletcher Public Library, Great Lakes Marine Collection, www.alpenalibrary.org /greatlakesmaritime

Ancestry, www.ancestry.com

Fulton History, www.fultonhistory.com

Maritime History of the Great Lakes, www.maritime historyofthegreatlakes.ca

Newspapers.com Publisher Extra, www.newspapers .com

NOAA Central Library, "Daily Weather Maps." n.d. US Department of Agriculture Weather Bureau. library.noaa.gov/Collections/Digital-Collections/US -Daily-Weather-Maps

NOAA Historic Maps and Charts Collection, https://historicalcharts.noaa.gov/historicals/search

Newspapers Cited

Alton (OH) Evening Telegraph, Bismarck (ND) Tribune, British Whig (Kingston, ON), *Buffalo (NY) Commercial Advertiser, Buffalo (NY) Daily Courier, Daily National Pilot* (Buffalo, NY), *Buffalo (NY) Daily Republic, Buffalo (NY) Journal and General Advertiser, Buffalo (NY) Morning Express, Buffalo (NY) Whig and Journal, Butte (MT) Daily Miner, Chicago Daily Tribune, Chicago Inter Ocean, Chicago Journal, Chicago Times, Daily True Democrat* (Cleveland, OH), *Cleveland (OH) Herald, Cleveland (OH) Morning Leader, Cleveland Plain Dealer, Cleveland Weekly Herald, Daily Chicago American, Daily Milwaukee News, Detroit Free Press, Door County Advocate, Erie (PA) Gazette, Fond Du Lac Weekly Union, Fort Wayne (IN) Daily Gazette, Grand River Times* (Grand Haven, MI), *Hunt's Merchants' Magazine* (New York, NY), *Manitowoc (WI) Pilot, Marine Record* (Cleveland, OH), *Marine Review* (Cleveland, OH), *Milwaukee (WI) Sentinel, Nevada State Journal* (Reno, NV), *New York Times, Niles' National Register* (Baltimore, PA), *Northern Advance* (Barrie, ON), *Oswego (NY) Palladium, Times-Herald* (Port Huron, MI), *Journal Times* (Racine, WI), *Racine Press, Sandusky (OH) Clarion, Sturgeon Bay (WI) Advocate, Yale (MI) Expositor.*

Notes

Short-form references are used in the notes to cite all sources, and full entries can be found in the bibliography.

Introduction

1. Church, *Life of John Ericsson*, 109–10.

Chapter One

1. Games, "Introductions, Definitions, and Historiography," 3–7. The new Atlantic history and Atlantic World scholarship is vast but has exercised only a limited influence on Great Lakes maritime historical scholarship. An important recent exception is Cangany, *Frontier Seaport*.
2. Labaree et al., *America and the Sea*.
3. This historical overview is based the author's overview essay "The Great Lakes," in Hattendorf, *Oxford Encyclopedia of Maritime History*, 105–11.
4. Parry, *Discovery of the Sea*, xii.
5. Cooper, *Survey of Submerged Cultural Resources*, 9–10.
6. Thomson, *Structures of Change*, 39–40. The most authoritative book on War of 1812 freshwater navy shipbuilding combines exhaustive archaeological and historical analysis; see Crisman, *Coffins of the Brave*.
7. Thiesen, *Industrializing American Shipbuilding*, 49.
8. Lewis, "The Frontenac."
9. Alden, "Old Shipbuilders of New York," 223–41.
10. Crisman, *Coffins of the Brave*, 363; Morrison, *History of New York Shipyards*, 39–40; Marestier, *Memoir on Steamboats*, 74.
11. Palmer, "First Steamboat on the Great Lakes," 7–8.
12. *Buffalo Daily Courier*, 3 June 1859. Transcribed at Maritime History of the Great Lakes (MHGL), http://images.maritimehistoryofthegreatlakes.ca/29514/data?n=5.
13. Crisman, *Coffins of the Brave*, 322–29.
14. Walker, *Early Days on the Lakes*, 302–3.
15. *Cleveland Weekly Herald*, 17 July 1823.
16. *Sandusky Clarion*, 17 July 1823.
17. Taylor, *Transportation Revolution*, 44–48; Cronon, *Nature's Metropolis*, 64.
18. Smith, *History of Wisconsin*, 466.
19. Discussions of the history of the state seal are derived from Holzheuter, "Wisconsin's Flag," 90–121.
20. Dana, *Two Years Before the Mast*, 1.

Chapter Two

1. Morris, *Miscellany*, 370–71.
2. For additional details on the environmental and business sides of early passenger steamships, see Hilton, *Lake Michigan Passenger Steamers*. A more recent volume that draws on Wisconsin Historical Society's earlier historical and archaeological documentation of the *Niagara* is Stone, *Floating Places*. The interpretation of the palace steamer era and of the *Niagara* as presented here is an extension of that published in Jensen, "History and Archaeology," 198–230.
3. Barton, *Brief Sketch of the Commerce*, 6.
4. Hall, *The West*, 10.
5. Taylor, *Transportation Revolution*. The major book on inland steam navigation remains Hunter, *Steamboats on the Western Rivers*.
6. Whitman, *Nelson's Biographical Dictionary*, 543.
7. Hezekiah Niles, *Niles' Weekly Register* 35 (7 October 1826); Sanford, *History of Erie County*, 135.
8. "Enrollment at Port Buffalo"; Historical Society of Pennsylvania, *Encyclopedia of Contemporary Biography of Pennsylvania*.
9. *Buffalo Journal and Advertiser*, 14 November 1832; *Buffalo Whig & Journal*, 5 September 1835.
10. *Hunt's Merchants Magazine and Commercial Review*, Vol. VI (1842): 442.
11. *Cleveland Weekly Herald*, 28 May 1835; *Buffalo Whig*, 15 April, 26 November 1834.
12. Ancestry.com, *Pennsylvania Biographical Sketches*.
13. Railroad Convention, *Proceedings*, 19; Reeves, "Navigation on Lake Erie," 308–10; "General Charles M. Reed," 404.
14. Stevenson, *Sketches*, 48.
15. Stevenson, *Sketches*, 51.
16. Stevenson, *Sketches*, 58, 65–68.
17. Stevenson, *Sketches*, 117.
18. Stevenson, *Sketches*, 58.
19. Stevenson, *Sketches*, 123, 150, 157,118.
20. *Buffalo Commercial Advertiser*, 19 May 1837.

21. Stevenson, *Sketches*, 63, 123, 150, 157,118;
22. Stevenson, *Sketches*, 57; Reeves, "Navigation on Lake Erie," 308–10.
23. Sanford, *The History of Erie County*, 118, 124–5, 128, 200–10.
24. "General Charles M. Reed," 400–10; *Nelson's Biographical Dictionary*, 543.
25. *Buffalo Commercial*, 11 May 1839; Reeves, "Navigation on Lake Erie," 195; *Hunt's Merchants Magazine*, Vol. VI (1842): 443.
26. *Daily Chicago American*, 20 April 1839.
27. "General Charles M. Reed," 400–10.

Chapter Three

1. The recycling of steam engines in new steamers during the 1840s conclusively demonstrated the superior efficiency of longer hulls. Whitaker, "Influence of Model," 102–4.
2. *Buffalo Commercial*, 23 August 1844. Transcribed at Maritime History of the Great Lakes (MHGL), http://images.maritimehistoryofthegreatlakes.ca/31585/data?n=4.
3. Ibid.
4. *Buffalo Commercial*, 29 May 1844; 23 August 1844. Transcribed at MHGL, http://images.maritime historyofthegreatlakes.ca/31585/data?n=4.
5. Whitaker, "Influence of Model," 102.
6. Walker, *Early Days on the Lakes*, 292.
7. Buffalo *Daily National Pilot*, 5 June 1845.
8. *Journal of the Franklin Institute* 12: 2 (1846).
9. Abbot, "Novelty Works," 721–34.
10. Coombe, "James P. Allaire," 263–71; Coombe, "Life and Times of James P. Allaire."
11. Russell, "On American River Steamers," 114.
12. Russell, "On American River Steamers," 114.
13. Haswell, *Engineers' and Mechanics' Pocket-Book*, 605.
14. Credit for most of the engine description goes to archaeologist Frank J. Cantelas, who led the underwater machinery documentation. Any errors in interpretation that may have crept in are my own.
15. *Niles National Register*, 20 December 1848.
16. *Buffalo Commercial*, 10 January 1846. Transcribed at MHGL, http://images.maritimehistoryofthe greatlakes.ca/29730/data?n=3.
17. Buckingham, *Eastern and Western States*, 268.
18. James, *Rambles*, 109.
19. Marjoribanks, *Travels*, 276.
20. Bird, *Englishwoman in America*, 168.

Chapter Four

1. Buffalo *Daily National Pilot*, 16 April 1846; Buffalo *National Pilot*, 12 May1845 and 16 April 1846; *Buffalo Daily Courier*, 16 April 1846.
2. *Buffalo Commercial*, 25 April 1846.
3. *Buffalo Commercial*, 5 May 1846; Odle, "Great Lakes History," 10.
4. *Buffalo Commercial*, 20, 28 May 1846; *Buffalo Daily News Pilot* 28 May 1846.
5. *Buffalo Daily News Pilot*, 10 June 1846.
6. *Milwaukee Sentinel*, 18 June, 24 July 1846.
7. *Milwaukee Sentinel*, 30 July1846.
8. *Cleveland Herald*, 27 July 1846.
9. *Buffalo Commercial*, 15 October 1846.
10. Excerpt from the Statement of T. Richards and Wm. Davenport sworn before Charles M. O'Mally, Justice of the Peace, Michiumakcinac [*sic*], 23 October1846. Transcribed copy provided by Louis and Janet Tuttle, Reston, Virginia, personal communication, January 30, 1996.
11. Thomas Richards to Louisa Richards, 25 October 1846. Transcribed by Louis and Janet Tuttle, Reston, Virginia, personal communication, 8 February 1998.
12. Louis and Janet Tuttle, miscellaneous research notes; *Erie Gazette*, 19 July 1849.
13. *Buffalo Commercial*, 23 November 1846.
14. *Buffalo Daily Courier*, 24 November 1846.
15. *Buffalo Commercial Advertiser*, 18 November 1846.
16. *Buffalo Morning Express*, 18 March 1847; 25 January 1848; 19 April 1849; 10 April 1850.
17. *Buffalo Morning Express*, 18 July 1847.
18. *Cleveland Plain Dealer*, 21 July 1847.
19. *Buffalo Morning Express*, 4 September 1847.
20. Marley, *Life of Reverend Thomas A. Morris*, 9; Morris, *Miscellany*, 361–71.
21. Morris, *Miscellany*, 367.
22. Morris, *Miscellany*, 367.
23. Morris, *Miscellany*, 369.
24. Morris, *Miscellany*, 367.
25. Spencer, "Reed Mansion."
26. *Buffalo Commercial*, 22 August 1844.
27. Wade, *Urban Frontier*.
28. Morris, *Miscellany*, 366–7.
29. Morris, *Miscellany*, 368.
30. Errols, "Thurlow Weed's Report," 167.
31. MacKinnon, *Atlantic and Transatlantic*, 117.
32. Bird, *Englishwoman in America*, 171.
33. Seibert, *Underground Railroad*, 82–3.

34. Morris, *Miscellany*, 366.

35. Morris, *Miscellany*, 369.

36. Morris, *Miscellany*, 370–1

37. *Buffalo Commercial*, 18 May 1848.

38. Odle, "Great Lakes History," 12–15; Williams, "Chicago River," 624.

39. *Buffalo Commercial*, 29 July 1848.

40. *Cleveland Daily True Democrat*, 3 August 1848.

41. Mansfield, *History of the Great Lakes*, 659.

42. *Buffalo Courier*, 14 April 1849.

43. *Milwaukee Sentinel*, 18 August 1849.

44. *Cleveland Daily True Democrat*, 15 August 1849; *Milwaukee Sentinel*, 21 August 1849.

45. *Buffalo Morning Express*, 3 September 1849.

46. *Buffalo Commercial*, 7 November 1849.

47. C. Patrick Labadie, Miscellaneous undated research notes from the collection of marine historian C. Patrick Labadie, Superior, Wisconsin.

48. *Buffalo Morning Express*, 18 February 1851.

49. *Buffalo Morning Express*, 16 May 1851.

50. *Buffalo Morning Express*, 21 July 1851.

51. *Buffalo Morning Express*, 1 December 1851.

52. Odle, "Great Lakes History," 13.

53. *Cleveland Morning Leader*, 24 April 1854; *Chicago Journal*, 1 April 1854.

54. *Cleveland Morning Leader*, 22 July 1854.

55. C. Patrick Labadie, Miscellaneous undated research notes from the collection of marine historian C. Patrick Labadie, Superior, Wisconsin.

56. Smith, "Northern Railway," 35–37.

57. *Milwaukee Sentinel*, 10 May 1856.

58. *Chicago Daily True Democrat*, 23 September 1856.

Chapter Five

1. Paskoff, *Troubled Waters*, 257.

2. *Acts of Congress Relating to Steamboats.*

3. *Buffalo Daily Republic*, 19 July 1856.

4. Passengers identified though the search of multiple contemporary newspaper accounts.

5. "Statement of Captain Miller," reprinted in *Weekly Wisconsin*, 26 September 1856.

6. "Statement of Captain Miller."

7. McDonald, "Eyewitness Reports," 195–202.

8. Hardridge, *Modern American Spiritualism*, 263.

9. McDonald, "Eyewitness Reports," 195–202; *Weekly Wisconsin*, 9 September 1856.

10. *Milwaukee Sentinel*, 6 September 1856; *Buffalo Commercial*, 29 September 1856.

11. *Milwaukee Sentinel*, 29 September 1856.

12. *Fond du Lac Weekly Union*, 2 October 1856.

13. *Fond du Lac Weekly Union*, 2 October 1856.

14. Act to Limit the Liability of Ship-owners; see *American Transportation Company v Moore*, Michigan Supreme Court, July 1858, for more details.

15. James, *Rambles*, 109.

16. Lane, *American Paddle Steamboats*; Shipley and Addis, *Paddlewheelers*; Stanton, *American Steam Vessels*.

17. Russell, "On American River Steamers," 114.

18. Russell, "On American River Steamers," 112–113.

19. Olsen, *Lady Elgin*, 9.

20. Baskerville, "Donald Bethune's Steamboat Business"; Cantelas, "Archaeological Investigation."

21. *U.S. Nautical Magazine and Naval Journal*, Vol. 5 (Dec. 1856): 235.

22. *U.S. Nautical Magazine*, Vol. 6 (June 1857): 192.

23. Reeves, "Navigation on Lake Erie," 23.

24. Herdendorf and Schuessler, *First Half Century of Merchant Steam.*

25. Lenihan, *Shipwrecks*, 128–34.

26. Lenihan, *Shipwrecks*, 37–40.

Chapter Six

1. Blois, *Gazetteer of the State of Michigan*; Williams, "Products of the Forest," 17–18.

2. Karamanski, *Schooner Passages*, 69; Cronon, *Nature's Metropolis*, 172.

3. Tunell, "Transportation on the Great Lakes," 106.

4. American Board of Commissioners for Foreign Missions, *Report*, 140.

5. Schoolcraft, *Personal Memoirs*, 471–72.

6. Lillie, *Historic Grand Haven*, 124–30.

7. Lillie, *Historic Grand Haven*, 97–100.

8. Lillie, *Historic Grand Haven*, 129.

9. Lillie, *Historic Grand Haven*, 186.

10. Karamanski, *Schooner Passages*, 65; Mansfield, *History of the Great Lakes*, 514.

11. Hotchkiss, *Industrial Chicago*, 40.

12. Hotchkiss, *Industrial Chicago*, 45–46.

13. Hotchkiss, *Industrial Chicago*, 679; Cronon, *Nature's Metropolis*, 169–171.

14. Cronon, *Nature's Metropolis*, 151; Dreyful, "City Transformed."

15. Griffiths, *Treatise*, 448.

16. Bates, "The Tables of the Schooner Challenge," 106–111.

17. *Grand River Times*, 31 August 1853.

18. *Grand River Times*, 26 October 1853.

19. Bates, "Mould-Loft Tables of the Schooner Magic," 498–501.

20. Findlay, *Dwight L. Moody*, 96.

21. Meverden and Thomsen, *Wisconsin's Cross-Planked Mosquito Fleet*.

22. Cronon, *Nature's Metropolis*, 152.

23. Bajema, "Blendon Landing," 107.

24. US Census Bureau, *Ottawa County, Michigan, Population Schedule*.

25. *Grand Haven News*, 19 June 1861.

26. *Detroit Free Press*, 2 October 1861.

27. For example, the *Ocean Wave*, a two-masted scow schooner built at Harsens Island, Michigan, also had a registered depth of 7 feet. Meverden and Thomsen, *Cross-Planked Mosquito Feet*, 30.

28. Meverden and Thomsen, *Myths and Mysteries*, 10.

29. Barkhausen, *Focusing on the Centerboard*, 18.

30. *Oswego Daily Times*, 19 March 1852; and Bates, "Inland Navigation," 15. In June 2018, Wisconsin Historical Society underwater archaeologists identified the *Montgomery* as the sixth and largest two-centerboard schooner shipwreck in the state. The *Montgomery* wrecked near Sheboygan, Wisconsin, in 1890.

31. Milwaukee *Weekly Wisconsin*, 5 April 1848; 5 October 1853; *Buffalo Daily Courier*, 9 April 1863; *Buffalo Commercial*, 7 September 1867.

32. *Buffalo Daily Courier*, 30 March 1852; *Racine Journal*, 17 July 1867.

33. Allen, *Sebago Lakes Region*.

34. Bates, "Marine Insurance."

35. Bates, "Inland Navigation," 15.

36. *Grand Haven News*, 16 July, 6 August 1862; *Lumberman* Certificate of Enrollment Detroit District issued August 8, 1862; *Chicago Tribune*, 15 August 1862.

37. Most of the information for this analysis is drawn from the *Chicago Tribune* and *Chicago Inter Ocean* newspapers.

38. *Marine Review* 9 (April 1909), 29.

39. *Chicago Tribune*, 11, 13 October 1869; 14 November 1881; *Chicago Inter Ocean*, 16 June 1874; 22 October 1878; *Detroit Free Press*, 24 October 1878.

40. *Chicago Tribune*, 6 December 1867.

41. *Chicago Tribune*, 27 September 1880; *Chicago Inter Ocean*, 5 June 1882; *Chicago Tribune*, 5 September 1882; *Detroit Free Press*, 20 November 1879.

42. *Marine Record*, 30 August 1883.

43. *Richmond Dispatch,* 3 March 1864.

44. *Grand Traverse Herald*, 29 July 1871.

45. McKay 1928: 332; *Grand Traverse Herald*, 29 April 1869; (Michigan. Northport County, 1870 U.S. Census: population schedule. Digital Image, Ancestry.com)

46. Detroit Enrollment No. 162, August 8, 1863.

47. Grand Haven Enrollments No. 365, December 5, 1865: No. 47, April 5, 1867; No. 53, March 26, 1869.

48. *Fort Wayne Daily Gazette*, 22 May 1881; *Senate Executive Journal*, 22 May 1871, p. 98; *Butte Daily Miner*, 17 February 1883; *Nevada State Journal*, 16 May 1886.

49. *Bismarck Tribune*, 9 February 1881.

50. Hales and Brimhall, "William and Jeannette Ferry," 128.

51. Chicago Enrollment No. 53, March 30, 1883; No. 89, April 24, 1884; Milwaukee Enrollment No. 41, March 23, 1886; No. 63, February 18, 1893.

52. Karamanski, *Schooner Passage*, 72–73; *Manitowoc Pilot*, 9 January 1890.

53. *Door County Advocate*, 6, 24 June 1893; 1, 7 July, 1893; 26 August 1893.

54. *Racine Journal Times*, 8 April 1893.

55. Tunell, *Transportation*, 97.

56. Cronon, *Nature's Metropolis*, 151.

57. Cronon, *Natures Metropolis*, 154.

Chapter Seven

1. The history presented in this chapter is an expansion of the author's work contributed to Meverden, Thomsen, and Jensen, *Wheat Chaff and Coal Dust*.

2. 1860 US Census Population Schedule, New York, Erie County; 1870 US Census Population Schedule, New York, Erie County.

3. Figures compiled from Heath, *Early Memories of Saugatuck*, 79–81; Milwaukee Public Library Ship Files Database, http://old.mpl.org/ship.

4. Heath, *Early Memories of Saugatuck*; 1900 U.S. Census Schedule, Michigan, Saugatuck Township.

5. Thomsen, Zant, and Kiefer, *Great Lakes Sailing Canallers*, 9; Labadie, *Pictured Rocks*, 21.

6. *Kate Kelly* Permanent Enrollment, Buffalo, May 23, 1867, no number available.

7. *Buffalo Daily Courier*, 23 March 1867.

8. *Kate Kelly* Vessel Enrollments, no. 126, 306, 18; *Daily Milwaukee News*, 24 October 1868; 19 April 1868.

9. *Detroit Free Press*, 20 May 1869.

10. Captain J. W. Hall (1869). Transcribed at Maritime History of the Great Lakes (MHGL), http://images .maritimehistoryofthegreatlakes.ca/46063/data.

11. Board of Lake Underwriters, *Classification of Lake Vessels.*

12. *Buffalo Commercial*, 10 August 1871.

13. *Chicago Inter Ocean*, 25 December 1874. Transcribed at MHGL, http://images.maritimehistory ofthegreatlakes.ca/48477/data?n=1; *Chicago Tribune* 10 September 1874.

14. *Oswego Palladium*, 27 September 1875. Transcribed at MHGL, http://images.maritimehistoryofthe greatlakes.ca/57213/data?n=1.

15. *Oswego Palladium*, 29, 30 September 1875, 4 October 1875; *Buffalo Commercial*, 1 October 1875.

16. *Oswego Palladium*, 6, 9, 12, 25 October 1875; *Oswego* Enrollment no. 130, 16 April, 1873.

17. North District of Ohio, Admiralty Case Nos. 1515, 1527, 1531, 1532, 1533, 1534, 1535.

18. North District of Ohio, Admiralty Case No. 1515.

19. North District of Ohio, Admiralty Case No. 1515.

20. North District of Ohio, Admiralty Case No. 1532.

21. Alford, "Shipbuilding Days," 88–94, 219–25, 306–12.

22. *Oswego Daily Times*, 5 August 1886.

23. *British Whig*, 25 June 1886; *Detroit Free Press*, 12 November 1891.

24. *British Whig*, 16 October 1890; *British Whig*, 21 May 1892; *Oswego Palladium*, 4 August 1886.

25. Estimates based on trip data reported in the *British Whig*, 16 June 1883, 23, 26, 28 July 1883; 21 August 1883, 1 September 1883, 28 May 1888, 1, 6, 18 June 1888; *Buffalo Morning Express*, 3 June 1891.

26. *Chicago Inter Ocean*, 19 May 1895.

27. Alton *Evening Telegraph*, 31 May 1878.

28. *Boston Post*, 26 October 1878; *Detroit Free Press*, 30 August 1879.

29. Thomsen et al., *With Waves in Their Wake*, 25–26.

30. The *Mary L. Higgie* was sold and renamed the *Hattie A. Estelle*, and it wrecked at Manistee, Michigan, with the loss of three lives in 1891.

31. *Chicago Inter Ocean*, 15 May 1895.

32. Bowlus, *Iron Ore Transport*, 157–8.

33. Mansfield, *History of the Great Lakes*, 759; *British Whig* 16, 18 June 1891, 22 July 1891; New York *Sun*, 14 August 1891; *Buffalo Enquirer*, 18 December 1891.

34. *Detroit Free Press*, 15 May 1895.

35. *Door County Advocate*, 28 October 1893; *Buffalo Daily Courier*, 2 November 1893.

36. *Buffalo Daily Courier*, 2 May 1894.

37. Weather information derived from US Department of Agriculture Weather Bureau daily maps. Available through the NOAA Central Library digital collections. www.lib.noaa.gov/collections/imgdocmaps /daily_weather_maps.html.

38. *Chicago Daily Tribune*, 12 May 1895.

39. *Chicago Inter Ocean*, 14 May 1895.

40. *Chicago Inter Ocean*, 14, 19 May 1895.

41. *Racine Journal*, 13 June 1895.

42. *Detroit Free Press*, 15 May, 16 May 1895, 12 June 1895; *Weekly Wisconsin*, 25 July 1895.

43. *Chicago Inter Ocean*, 19 May 1896.

44. *Chicago Inter Ocean*, 19 May 1895.

45. The site description comes the information published in Meverden, Thomson, and Jensen, *Wheat Chaff and Coal Dust.* The author did not take part in the underwater archaeological documentation of the wreck site.

46. Meverden, Thomsen, and Jensen, *Wheat Chaff and Coal Dust*, 43.

47. US Patent Office, *Specifications of Patents*, 377–378.

Chapter Eight

1. Mansfield, *History of the Great Lakes*, 973.

2. Havighurst, *Long Ships Passing*, 262.

3. Quinney, *Bryant College*, 9–27.

4. *Buffalo Daily Courier*, 4 April 1856.

5. Plumb, "1857 Depression on the Lakes," 290; Mansfield, *History of the Great Lakes,* 677–9.

6. Albion, *Square Riggers on Schedule*, 174–201.

7. Melville, quoted in Maddox, *The Atlantic Crossing*, 90

8. Albion, *Square-Riggers on Schedule*, 151.

9. Albion, *Square-Riggers on Schedule*; Maddocks, *The Atlantic Crossing*, 102.

10. Maddocks, *The Atlantic Crossing*, 106–111.

11. Howe and Mathews, *American Clipper Ships.*

12. *New York Daily Herald*, 6 September 1862; "Clipper Panther."

13. *Detroit Free Press*, 6 February 1929.

14. *Buffalo City Directory*; Havighurst, *Long Ships Passing*, 262.

15. Havighurst, *Long Ships Passing*, 262; Roberts, personal communication.

16. Havighurst, *Long Ships Passing*, 262.

17. *Lake Vessel Register*, 15; Bureau of Navigation, *List of Merchant Vessels*; Baker, *Shipbuilding on the*

Saginaw, 19–20; C. Patrick Labadie, 1992–93 correspondence with author, materials on file at the State Historical Society of Wisconsin, Historic Preservation Division, Madison, Wisconsin. Hereafter referenced as Labadie, personal communication.

18. Mansfield, *History of the Great Lakes*, 253.

19. These figures are derived from the 1874 Board of Lake Underwriters vessel list and offer a comparison of relative activity during this period rather than firm numbers. Other schooners were built but not included on the list.

20. *Chicago Daily Tribune*, 14 April 1872.

21. *Chicago Daily Tribune*, 14 May 1872.

22. Baker, *Shipbuilding on the Saginaw*, 1.

23. *Buffalo Commercial*, 17 September 1872.

24. Kilar, *Michigan's Lumbertowns*, 21–22.

25. *Buffalo Commercial*, 17 September 1872.

26. *Future Prospects of Bay City*, 63; *History of Bay County*, 50, 166, 223–25.

27. Baker, *Shipbuilding on the Saginaw*, 2.

28. Thompson, *Steamboats and Sailors*, 33–34.

29. Lenihan, *Submerged Cultural Resource Study*, 57–58.

30. *Buffalo Morning Express*, 8 August 1874.

31. "Samuel F. Hodge," 649–53.

32. *Buffalo Commercial*, 10 August 1874.

33. *Port Huron Times-Herald*, 18 April 1874; *Buffalo Morning Express*, 21 July 1874.

34. *Buffalo Commercial Advertiser*, 10 August 1874; *Chicago Inter Ocean*, 31 August 1874.

35. Detroit Free Press, 16 July 1874; Oswego Palladium, 28 June 1874.

36. Rae, "Great Lakes Commodity Trade," 256; McCarty, "Economic Aspects," 142–43; Mansfield, *History of the Great Lakes*, 535.

37. Williamson, "Growth of the Great Lakes," 188–89.

38. *Chicago Inter Ocean*, 16 September 1874.

39. *Buffalo Daily Courier*, 17 September 1874; *Chicago Inter Ocean*, 19 September 1874; 11 October 1874.

40. Karamanski, *Schooner Passage*, 130–31.

41. *Chicago Daily Tribune*, 12 November 1874.

42. *Chicago Daily Tribune*, 19 November 1874.

43. *Chicago Inter Ocean*, 2 September 1878.

44. *Chicago Daily Tribune*, 11 July 1877; 3 September 1877.

45. Estimates are based on freight rates and deliveries reported in the marine columns of various newspapers.

46. *Chicago Inter Ocean*, 12, 26 November 1877; *Chicago Daily Tribune*, 3 December 1877.

47. *Chicago Tribune*, 8 August, *Chicago Inter Ocean*, 19 September 1879; *Chicago Tribune*, 27 May 1880.

48. Baker, *Shipbuilding on the Saginaw*, 1; Roberts, personal communication; *Detroit Free Press*, 24 March 1880.

49. Hall, *Report on the Ship-Building Industry*, 140; Bureau of Navigation, *List of Merchant Vessels*; Dickson, "Longest Schooner," 2–6.

50. Hall, *Report on the Ship-Building Industry*, 140; Wood, "Morning with Valentine Fries," 172–74; Dickson, "Longest Schooner," 8–9.

51. Bureau of Navigation, *List of Merchant Vessels*; Wright, *Freshwater Whales*, 24.

52. *Chicago Inter Ocean*, 6 December 1880.

53. Hall, *Report on the Ship-Building Industry*, 172.

54. Hall, *Report on the Ship-Building Industry*, 172.

55. Hall, *Report on the Ship-Building Industry*, 172–73; Hall, "Ship Building in the United States," 75.

56. Hall Notebook, 76.

57. *Buffalo Commercial*, 30 September 1879; *Buffalo Evening News*, 24 March 1883.

58. Hall, *Report on the Ship-Building Industry*, 173.

59. Hall, *Report on the Ship-Building Industry*, 173.

60. *Detroit Free Press*, 31 July 1881

61. *Cleveland Herald*, 12 September 1881

62. *Chicago Tribune*, 5 September 1881; *Detroit Free Press*, 10 October 1881.

63. *Siberia* Port Huron Temporary Certificate of Enrollment No. 5, July 8, 1882; *Detroit Free Press*, 29 June 1882; *Chicago Inter Ocean*, 27 June 1882.

64. *Detroit Free Press*, 2 August 1882; *Chicago Inter Ocean*, 8 August 1882, 2 September 1882.

65. *Buffalo Daily Courier*, 21 August 1882.

66. *Buffalo Morning Express*, 13 September 1882.

67. *Chicago Inter Ocean*, 13 September 1882.

68. *Siberia* Buffalo Certificate of Enrollment No. 11, August 13, 1883; *Buffalo Daily Courier*, 18 November 1884.

69. *Marine Record*, 11 October 1883; *Buffalo Daily Courier*, 18 November 1884 (see Alpena County George Fletcher Library Great Lakes Maritime Collection for interior photograph of dining room and ship painting).

70. "*Australasia* National Register Nomination," 2013.

71. *History of Bay County, Michigan*, 225.

72. "*Australasia* National Register Nomination," 2013.

73. Bureau of Navigation, *List of Merchant Vessels*; Van der Linden, *Great Lakes Ships We Remember*, 355.

74. *Marine Record*, 28 May 1885; 12 November 1885.

Chapter Nine

1. Vessel statistics taken from Tunell, *Transportation on the Great Lake of North America*. H. doc. 277. 55th Cong. 2nd Sess. 1898.

2. Labadie, *Submerged Cultural Resources Study*, 10; Hogan, *Economic History*, 20.

3. Hogan, *Economic History*, 196–200.

4. Tunell, *Transportation on the Great Lake of North America*, 72–77.

5. Tunell, *Transportation on the Great Lake of North America*, 81.

6. Tunell, *Transportation on the Great Lake of North America*, 5.

7. Thompson, *Graveyard of the Lakes*, 92; Bowlus, *Iron Ore Transport*, 109–10.

8. Chernow, *Titan*; Mansfield, *History of the Great Lakes*, 364–7, 1067–8.

9. *Cleveland Herald*, 21 June 1876.

10. *Cleveland Herald*, 21 October 1876.

11. *Buffalo Express*, 28 August 1877.

12. *Chicago Inter Ocean*, 20 September 1877.

13. J. W. Hall Scrapbook 1877–1878, Maritime History of the Great Lakes Digital Library, http://images .maritimehistoryofthegreatlakes.ca/56138/data?n=5.

14. Murphy, "Shipwrecks as Database," 65–90.

15. *Buffalo Commercial*, 23 February 1884.

16. Tunell, *Transportation on the Great Lake of North America*, 24, 78.

17. Earnings estimates derived from issues of Cleveland *Marine Record* and other newspaper sources.

18. *Chicago Inter Ocean*, 19 May 1886.

19. Rates taken from various newspaper reports and issues of the weekly Cleveland *Marine Record*.

20. *Chicago Daily Tribune*, 6, 7 September 1886; *Chicago Inter Ocean*, 30 August 1886.

21. *Chicago Daily Tribune*, 27 September 1886.

22. *Chicago Daily Tribune*, 16 October 1886.

23. The report of original historical and archaeological investigation of the *Lucerne* is found in Cooper et al. *By Fire, Storm, and Ice*, 31–57. Cooper's research provided the original historical framework and archaeological interpretation for this expanded account. The author of this book participated in the fieldwork as underwater archaeologist.; *Marine Record*, 9 December 1886.

24. The weather information is taken principally from the daily weather maps digitized at https://library .noaa.gov/Collections/Digital-Collections/US-Daily -Weather-Maps.

25. *Marine Record*, 9 December 1886.

26. *Marine Record*, 9 December 1886; *Detroit Free Press*, 20 November 1886; Minneapolis *Star Tribune*, 20 November 1886.

27. Cooper et al., *By Fire, Storm, and Ice*, 34–37.

28. *Marine Record*, 25 November 1886

29. Cooper et al., *By Fire, Storm, and Ice*, 31–57.

30. *Detroit Free Press*, 28 September 1886.

31. *Philadelphia Inquirer*, 10 September 1887; *Marine Record*, 15 September 1887.

32. NOAA Daily Historical Weather Maps, https:// library.noaa.gov/Collections/Digital-Collections /US-Daily-Weather-Maps.

33. *Marine Record*, 4 September 1884.

34. *Marine Record*, 26 April 1887; 3, 5 May 1887.

35. *Marine Record*, 19 May 1887.

36. *Chicago Inter Ocean*, 10 September 1887.

37. *Buffalo Courier*, 26 February 1923

38. *Marine Record*, 15 September 1887.

39. *Detroit Free Press*, 10 September 1887.

40. *Marine Review*, 20 October 1887.

41. *Detroit Free Press*, 11 June 1887.

42. *Chicago Tribune*, 26 October 1887.

43. *Detroit Free Press* 26 October 1887.

44. *Marine Record*, 19 January 1888.

45. *Marine Record*, 26 January 1888.

46. *Chicago Tribune*, 26 October 1887.

47. *Marine Review*, 31 January 1895.

48. *Marine Record*, 21 December 1899.

49. *Blue Book of American Shipping 1900*, 339–41.

50. *Marine Record*, 14 December 1899.

51. Bellamy, *Last Days of Cleveland*, 74–95.

52. *Buffalo Morning Express*, 19 July 1900.

53. *Buffalo Morning Express*, 25 April 1901.

54. *Buffalo Commercial Advertiser*, 31 July 1891.

55. *Buffalo Commercial Advertiser*, 31 July 1891, 24 February 1902.

56. An extensive obituary for Holmes printed in the *Maryland Historical Magazine* provides a long list of accomplishments for Holmes, some of them, such as commanding the clipper ship *Glory of the Seas* as well as the largest freighter on the Great Lakes, utterly spurious. That he enlisted in the Navy and died in service to his country, however, is not in doubt. "Joseph Holmes Obituary," 299–300.

Chapter Ten

1. *Detroit Free Press*, 3 September 1897.
2. Some of the historical and all of the archaeological data in this section is adapted from Meverden and Thomsen, *Wisconsin Coal Haulers*.
3. Meverden and Thomsen, *Wisconsin Coal Haulers*, 51.
4. Meverden and Thomsen, *Wisconsin Coal Haulers*, 52.
5. *Door County Advocate*, 24 October 1896.
6. *Door County Advocate*, 24 October 1896
7. *Door County Advocate*, 31 October 1896; 28 November 1896; 6 February 1897; 16 October 1897.
8. Site description adapted from Meverden and Thomsen, *Wisconsin Coal Haulers*, 2013.
9. The archaeological documentation and much of the specific history of the *City of Glasgow* is adapted from Rodgers, *Bones of a Bulk Carrier*.
10. US Census Office, *Report on Manufacturing Industries*, 550–55.
11. *Detroit Free Press*, 25 January 1891.
12. Rodgers, *Bones of a Bulk Carrier*, 9.
13. *Buffalo Courier*, 28 June 1893.
14. *Marine Record*, 3 October 1895.
15. *Sturgeon Bay Advocate*, 28 May 1908.
16. *Sturgeon Bay Advocate*, 4 June 1908.
17. This overview of the *Frank O'Connor* is adapted in part from Cooper and Jensen, *Davidson's Goliaths*.
18. *Chicago Inter Ocean*, 30 August 1892.
19. *Chicago Daily Tribune*, 20 September 1892.
20. *Detroit Free Press*, 9 March 1893.
21. *Detroit Free Press*, 11 March 1893.
22. *Detroit Free Press*, 4 November 1898. The information on the Gilchrist fleet comes from the *Gilchrist Family Papers*, Bentley Historical Library, University of Michigan.
23 *Gilchrist Family Papers*.
24. *Gilchrist Family Papers*.
25. Cooper and Jensen in *Davidson's Goliaths* provide a detailed account of the condition of the site in 1991 on which this brief description is based.
26. *Detroit Free Press*, 20 November 1893.
27. Hilton, *Great Lakes Car Ferries*, 186–98.
28. Hilton, *Great Lakes Car Ferries*, 229–35.
29. Oldham, "Shipbuilding and Transportation," 506.
30. *Detroit Free Press*, 5 October 1895.
31. *Chicago Inter Ocean*, 21 September 1895.
32. *Chicago Inter Ocean*, 9 December 1895.
33. *Detroit Free Press*, 31 March 1896.
34. *Detroit Free Press*, 26 July 1896.
35. *Detroit Free Press*, 14 July 1896.
36. *Yale Expositor*, 23 October 1896.
37. *Chicago Tribune*, 13 January 1897.
38. *Detroit Free Press*, 24 August 1897.
39. *Chicago Inter Ocean*, 17 December 1897.
40. *Detroit Free Press*, 4 January 1898; *Gilchrist Family Papers*.
41. *Chicago Tribune*, 5 February 1899.
42. *Chicago Inter Ocean*, 12 August 1899.
43. *Marine Review*, 31 August 1899.
44. *Detroit Free Press*, 4 August 1900.
45. U.S. Life-Saving Service, *Annual Report*; *U.S. Life-Saving Service Station Log*, Milwaukee, 15 November 1905 Record Group 26.4.2, Records Relating to Operations—Milwaukee. National Archives Chicago Branch; *Milwaukee Sentinel*, 2, 11 November 1905; *Door County Advocate*, 26 September1907.
46. The *Pretoria* section is expanded from the research by the author for Cooper and Jensen, *Davidson's Goliaths*. Some of the text is taken from that publication.
47. *Chicago Inter Ocean*, 21 May 1900.
48. *Detroit Free Press*, 26 May 1900.
49. *Detroit Free Press*, 14 July 1900; *Gilchrist Family Papers*.
50. *Detroit Free Press*, 14 September 1900.
51. *Buffalo Courier*, 25 November, 10 September 1900.
52. *Saginaw Courier-Herald*, 13 September 1900; *Buffalo Morning Express*, 14 September 1900; *Chicago Inter Ocean*, 13 September 1900.
53. *Marine Record* 8 May 1902.
54. Cooper et al., *By Fire, Storm, and Ice*, 118–23.
55. Cooper and Jensen, *Davidson's Goliaths*, 51–52.
56. Wolff, "The Shipwrecks of Lake Superior," 174–187; *Chicago Tribune*, 6 September 1905.
57. Schneider, C. F. *Michigan Section*, 4.
58. *Buffalo Courier*, 6 September 1905.

Epilogue

1. Marx and Lawrence, "National Register of Historic Places Registration Form: Paul Palmer—Shipwreck and Remains."
2. *Detroit Free Press*, 9 February 1905.
3. Bates, *American Marine*, 345.
4. Winthrop L. Marvin, *American Merchant Marine*, 406.
5. Martin, "Grand Haven Rig," 201.

6. Parker, *Great Coal Schooners*, 37.

7. Marvin, *American Merchant Marine*, 368.

8. Parker, *Great Coal Schooners*, 47; Runge Collection, vessel data sheets—Wisconsin Marine Historical Society, Milwaukee Public Library.

9. Parker, *Great Coal Schooners*, 43.

10. *Marine Engineering*, October 1890, 430.

11. Parker, *Great Coal Schooners*, 107.

12. Morris, *American Sailing Coasters*, 49; Parker, *Great Coal Schooners*, 41–44.

13. Swayze, "Giant Wooden Barges," 104–8.

INDEX

ABOUT THE AUTHOR

John Odin Jensen has studied North American maritime frontier shipwrecks from the Grand Banks of Newfoundland to the edges of the Bering Sea. Born into a Norwegian-American seafaring family in Alaska, he began his maritime career working alongside his father and brother in the commercial fisheries in the 1970s, a time and place where shipwreck and death at sea were an accepted part of life. As a former crab boat captain and shipwreck survivor, Jensen brings deep professional experience and personal sympathy to the study of the North American mariners, ships, and shipwrecks. His more than thirty years of Great Lakes experience began with a position as an engineer/deckhand aboard the University of Wisconsin–Milwaukee research vessel *Neeskay* and continued with many seasons surveying shipwrecks as a professional underwater archaeologist with the Wisconsin Historical Society. In addition to his early seagoing education, Jensen earned a BA in history from Lawrence University, an MA in maritime history and underwater archaeology from East Carolina University, and MS and PhD degrees in history from Carnegie Mellon University. He resides in Pensacola, Florida, with his wife, Karen Belmore, and is on the faculty of the department of history at the University of West Florida.